MONOGRAPHS ON SOIL AND RESOURCES SURVEY

General Editor: P. H. T. Beckett

MONOGRAPHS ON SOIL AND RESOURCES SURVEY

1. Aerial photography and remote sensing for soil survey
   L. P. White (out of print)
2. Quantitative and numerical methods in soil classification and survey
   R. Webster
3. Soil sampling and soil description
   J. M. Hodgson
4. Soil survey contracts and quality control
   S. Western
5. Soil classification for soil survey
   B. E. Butler
6. Soil and permafrost surveys in the Arctic and subarctic
   K. A. Linell and J. C. F. Tedrow
7. Land evaluation
   S. G. McRae and C. P. Burnham
8. Soil survey for engineering
   A. B. A. Brink, T. C. Partridge, and A. A. B. Williams
9. Ground and air survey for field scientists
   John Wright
10. Soil resource surveys for forestry
    K. W. G. Valentine
11. Soils and Quaternary geology: a handbook for field scientists
    J. A. Catt
12. Principles of geographical information systems for land resources assessment
    P. A. Burrough
13. Surveying derelict land
    E. M. Bridges

Surveying Derelict Land

E. M. BRIDGES

Senior Lecturer,
Department of Geography,
University College, Swansea

CLARENDON PRESS · OXFORD
1987

Oxford University Press, Walton Street, Oxford OX2 6DP
Oxford New York Toronto
Delhi Bombay Calcutta Madras Karachi
Petaling Jaya Singapore Hong Kong Tokyo
Nairobi Dar es Salaam Cape Town
Melbourne Auckland

and associated companies in
Beirut Berlin Ibadan Nicosia

Oxford is a trade mark of Oxford University Press

Published in the United States
by Oxford University Press, New York

British Library Cataloguing in Publication Data
Bridges, E. M.
Surveying derelict land.—(Monographs on soil and resources survey; 13).
1. Reclamation of land—Great Britain
2. Waste lands—Great Britain
I. Title II. Series
333.73 HD1671.G7
ISBN 0-19-854566-5

Library of Congress Cataloging in Publication Data
Data available

Set by Cotswold Typesetting Ltd, Cheltenham
Printed in Great Britain
at the University Printing House, Oxford
by David Stanford
Printer to the University

Preface

Surveying derelict land is essentially team-work, drawing upon the expertise of different scientists and engineers. This book assembles an interdisciplinary methodology for the assessment of derelict and contaminated land. It is written for planners, engineers, ecologists, and earth scientists who become involved in the process of rehabilitating land damaged by industrial processes. It extends the concepts of soil survey to an assessment of waste materials and contaminated soils found on industrial sites and complements *Soil surveys for engineering* published in this series (Brink *et al.* 1982).

The author is grateful to Dr P. H. T. Beckett for his invitation to contribute this book to the Monographs on Soil and Resources Survey series, for his help and guidance in the formative stages of its preparation, and for his editing of the manuscript. Appreciation is also expressed to the following individuals and organizations who provided information and source material: M. J. Beckett (Interdepartmental Committee on the Redevelopment of Contaminated Land); Dr W. O. Binns for information on restoration for forestry; Dr H. J. M. Bowen for use of Table 5.2; Dr R. D. F. Bromley for material relating to the Lower Swansea Valley; CEGB (Ash Disposal Site, Peterborough) for information relating to Peterborough; Dr I. S. Clark for estimation of land restoration; B. Daniel (Thamesdown Engineers Department) and members of the Thamesdown Planning Department for information about Swindon; the Department of the Environment for diagrams and tables from their publications; Dr M. F. Downing for check-list of features of derelict land; J. Elias (Welsh Development Agency) for information about government assistance for the redevelopment of derelict land; Dr N. Jackson (Robinson, Jones Partnership, Blackwood) for information relating to the Rising Sun Site; H. Jones (Swansea City Council Engineers Department) for information about restoration in the Lower Swansea Valley; G. Lowe for information about Thamesmead; Dr G. Macinko (Washington State University, Ellensberg) for information relating to dereliction and restoration in the USA; Dr S. G. McRae (Wye College, London) for information about gravel-quarry restoration; the Ordnance Survey for permission to use extracts from 1:25 000 maps; G. Padgett (West Glamorgan County Council Planning Department) for bibliographic assistance; R. A. Page (Welsh Office, Environmental Protection Unit) for bibliographic assistance; A. Platt for information concerning derelict railway land; Dr N. H. Saunders for information upon radioactivity; and M. A. Smith (Bostock Hill and Rigby, formerly Building Research Establishment, Watford) for material about Thamesmead and contaminated land problems and how to solve them. Many other people have contributed ideas or practical assistance which is gratefully acknowledged.

In the preparation of the manuscript, I am extremely grateful to Dr D. W. Hopkins (Honorary Research Fellow in the Department of Chemistry and formerly senior lecturer in Metallurgy, University College of Swansea) for reading and commenting on the text. Mrs G. Bridges and Mrs M. Owens (Word Processing Services) corrected and typed the manuscript; G. B. Lewis, T. Fearnside, and Miss M. Sizer (Cartographic Unit, University College, Swansea) drew the illustrations assisted by G. Llewellyn and A. Lloyd (Luton College of Higher Education). D. E. Price and A. F. Cutliffe (Photographic Unit, University College of Swansea) provided photographic assistance with diagrams and plates.

Published material is acknowledged in the text and the author expresses thanks and appreciation to the original authors and publishers.

Much of the manuscript for this volume was written while the author held a NATO–CCMS fellowship which provided financial assistance for travel and research—particularly into the use of remote sensing for mapping and monitoring contaminants on derelict land. The remote sensing was undertaken in collaboration with M. G. Coulson (University College of Swansea) at the Space Research Laboratories of RAE Farnborough, using data provided by the Natural Environment Research Council, the headquarters of which is situated on rehabilitated derelict railway land at Swindon.

Swansea E. M. B.
April 1986

Contents

1 Introduction

The landscape which surrounds us has been modified by successive generations, so our heritage is an amalgam of much that is fine and splendid, and also much that is unsightly and dangerous. Land in this second category, which has been spoiled by industrial activity and subsequently abandoned, perhaps because of changes in technology and world economic conditions that rendered processes and plant obsolete, is frequently described as derelict. It may have been contaminated by industrial processes and may now harbour wastes and toxic materials. This book is concerned with the survey of derelict and contaminated land as a stage in the process of land reclamation.

Building stone, gravel, sand, and cement are required for construction, and coal, oil, and gas are required to provide heat and electrical energy. Yet these resources all come from the earth and their extraction and the dumping of waste materials may ravage the countryside. Many countries now have strict regulations concerning mineral extraction and the dumping of waste; even so, some degradation of the environment is unavoidable.

Consequently, the extent of derelict and contaminated land continues to increase. Many countries are attempting to restore ravaged land for productive purposes, but in most cases more land is becoming derelict than is being restored. Nearly 45 700 ha of land in Britain are recorded as derelict, of which 34 300 ha are considered worthy of reclamation (DoE 1984). Urban areas contain 46 per cent of this derelict land, and 92 per cent of this is thought suitable for reclamation. Between 1 April 1974 and 31 March 1982, 17 000 ha in England were reclaimed with the aid of government grants. However, the total area of derelict land increased during the same period from 43 273 to 45 700 ha, mainly through changes in the industrial north-west of England, in the Midlands, and in dockland areas (Figs. 1.1 and 1.2). The situation in Wales (Fig. 1.3) was surveyed in 1971–72 (Welsh Council 1976).

Recently, city-centre decay has introduced a new facet of dereliction. Numerous European cities suffered extensive damage through bombing during the Second World War and many sites remained derelict until the rebuilding programmes of the 1950s. Those cities which escaped the effects of war had a legacy of industrial and domestic architecture which was nearing the end of its useful life. Since then, new industrial processes have required different types of accommodation, and people's expectations have led them to expect higher standards of housing. Consequently, both industry and people vacated inner city areas and encroached on productive agricultural land on the city fringe, leaving the inner cities with decaying housing and abandoned factory sites. Rehabilitation has been slow, but if encouraged would alleviate the pressure for expansion on periurban land. Once land is urbanized it loses its flexibility for diverse farming practices, forestry, water supply, or recreation.

Defining derelict and contaminated land

In Britain interest was first shown in overcoming the problems of dereliction when enlightened landowners planted trees on the spoil heaps from mining. Enterprising local authorities in the Midlands cleared small spoil heaps as part of unemployment relief schemes in the 1930s, but the problem was not tackled on a national scale until towards the end of the Second World War, when questions were asked in Parliament, and the then Minister of Town and Country Planning was instructed to investigate. One of the first results was a report on derelict land in parts of central England, commonly referred to as the Black Country (Beaver 1946). After discussion with the local planning officers, Beaver then defined derelict land as:

> Land which has been so damaged by extractive or other industrial processes or by any form of urban development that in default of special action it is unlikely to be effectively used again within reasonable time and may well be a public nuisance in the meanwhile.

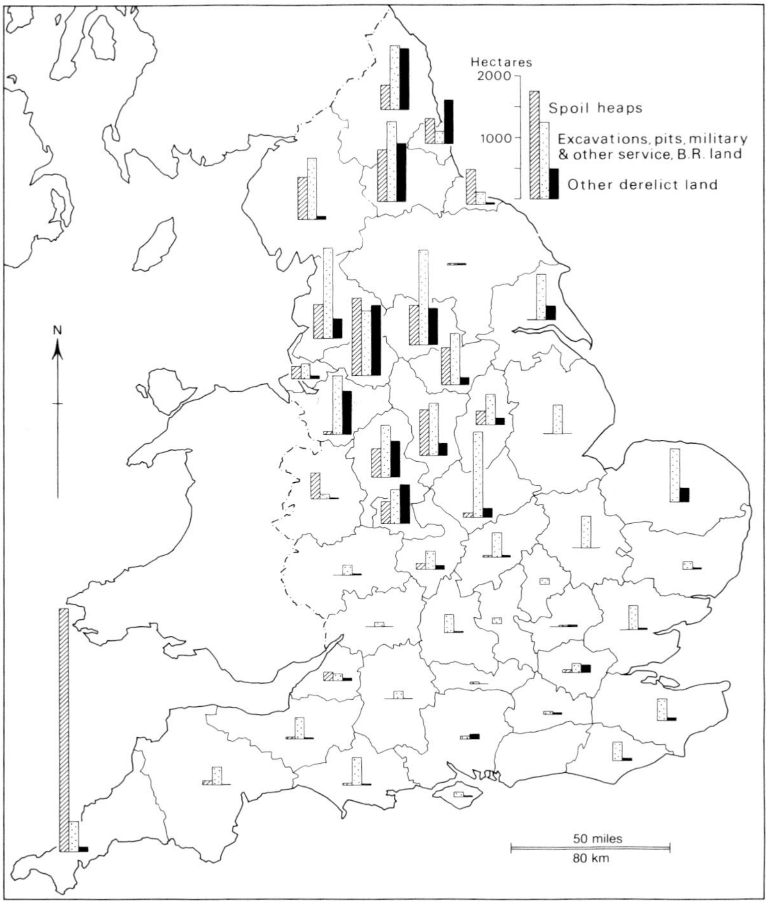

Fig. 1.1. Total derelict land in England, 1974.

In spite of this, there remains considerable confusion. The words degraded, despoiled, disturbed, devastated, damaged, and ravaged have all been applied to derelict land, but the strict dictionary definitions of these words do not always correspond to what the user intended. Although the word ravaged probably expresses the condition of this type of land most correctly, **derelict** is so well entrenched that its use has continued. The official definition of **derelict land** used in goverment publications (Ministry of Housing and Local Government 1966) is:

> Land so damaged by industrial and other development that it is incapable of beneficial use without treatment.

This is also called **disused derelict land**. The term **active derelict land** is used for damaged land which is still in use for tipping or other activities which create unsightly landscapes. Wallwork (1974) suggests the following terms:

Extant dereliction: land now derelict, as officially defined, but also including sites still covered by planning conditions and permission to extract minerals.

Potential dereliction: land that will, on cessation of present use, become derelict unless remedial action is taken, for example colliery waste tips and quarries.

Partial dereliction: may be described as either 'regressive' where land has been derelict but has been put to other uses without reclamation, or 'progressive' where deterioration has begun and the land will become derelict unless remedial action is taken.

For a variety of reasons certain categories of derelict land did not appear in official returns:

1. Active mineral extraction sites established after July 1948 normally have conditions which stipulate site restoration attached to the permission for extraction of the mineral. However, where extraction

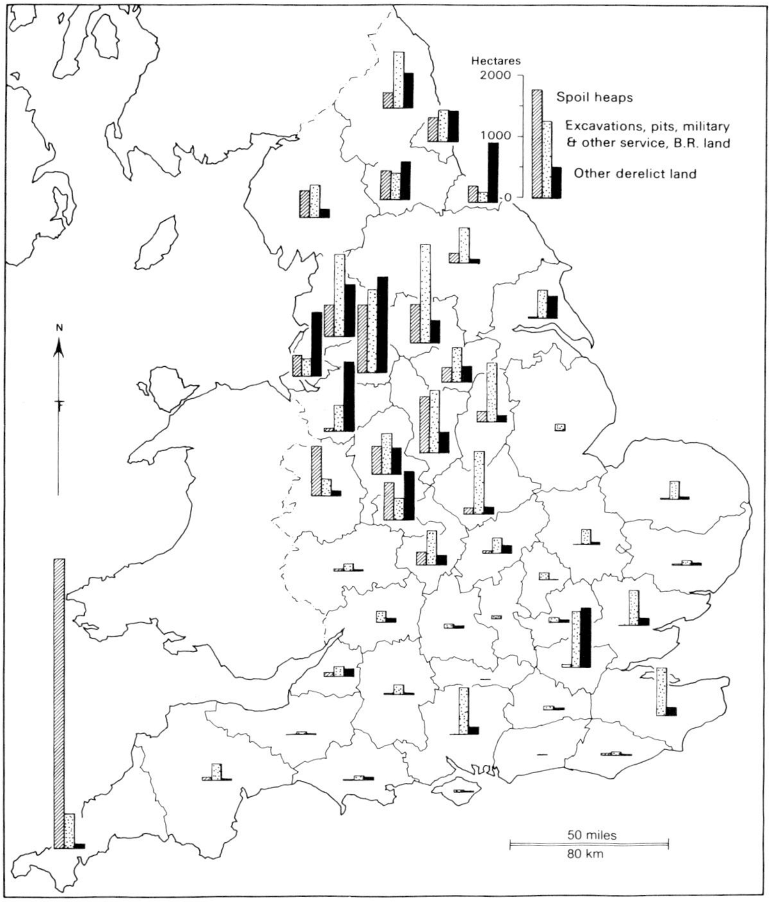

Fig. 1.2. Total derelict land in England, 1982.

has ceased for a long period, or if the extracting company goes into liquidation, such lands may remain in apparent dereliction for long periods. As they are not technically derelict, they are not included in the returns.

2. Successive legislation (general development orders) has excluded excavations and waste tips which were in existence in July 1948 from planning control. This effectively excludes about 25 per cent of all colliery waste tips from planning control and at the same time omits them from any category of derelict land in official statistics.

3. When permission for mineral workings in certain areas was given, in the immediate post-war period, extraction was allowed 'for the foreseeable future' and no specific limitation in years was stated. This means that such lands may also lie idle for long periods without extraction taking place, but as in previous cases they are not recorded as derelict.

4. Sites of major urban clearances often remain empty or in a derelict state for long periods, but presumably will eventually be rebuilt.

5. Land from which minerals have been extracted but which has undergone natural rehabilitation to such an extent that the site is aesthetically acceptable—for example, gravel pits now used for fishing or sailing, or quarries used for motorcycle scrambling.

6. Land may be damaged by fumes or by spillage of toxic chemicals, and/or may be flooded as a result of mining subsidence, so that it is virtually derelict.

Wallwork (1974) states that the last three categories are of less importance than the first three; together they add considerably to the total. A survey in the West Riding of Yorkshire in 1966 revealed

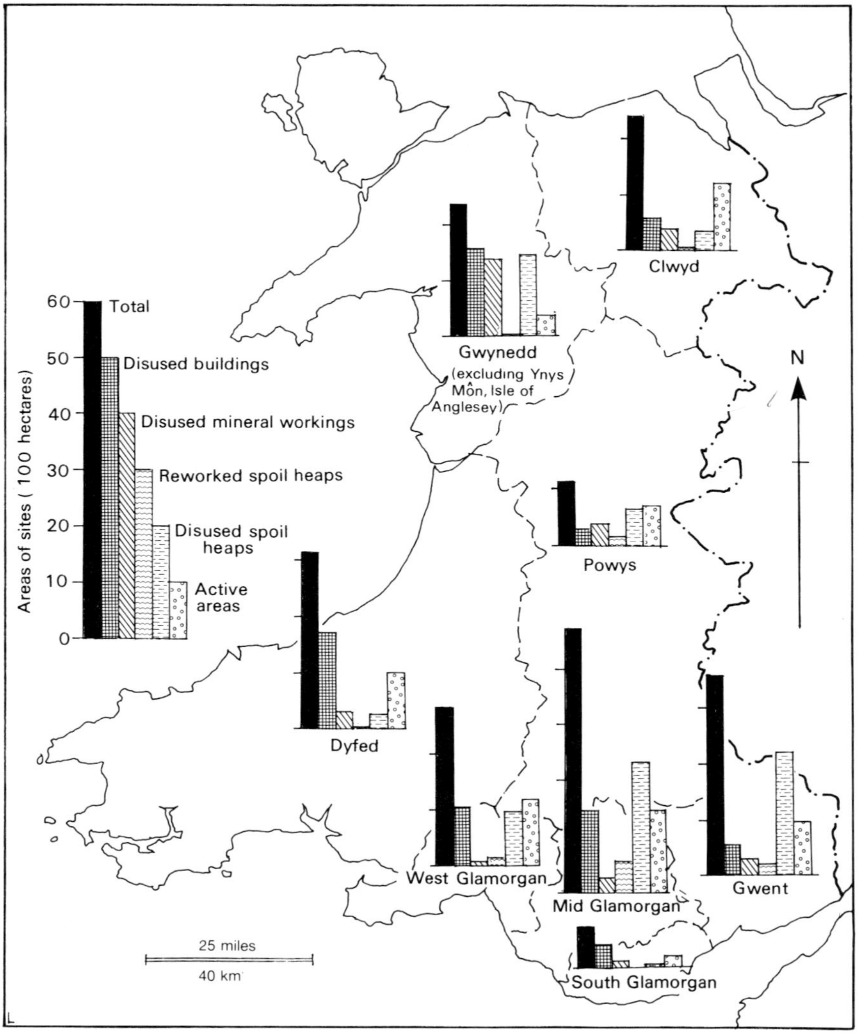

Fig. 1.3. Categories of derelict land in Wales, 1972.

6850 ha of derelict land compared with a mere 2540 ha in official figures, demonstrating the importance of a correct definition and interpretation of features on the ground.

Some derelict land was abandoned by its owners because the problems of returning it to productive use were insurmountable by the individual landowner or company. Many sites where operations were simply extractive may not be contaminated with toxic materials, but often derelict land has been contaminated. *Contaminated land* contains 'substances that when present in sufficient quantities or concentrations are likely to cause harm, directly or indirectly to man, to the environment, or on occasions to other targets' (Smith 1985). This definition of contaminated land embraces the presence of substances that may be harmful to plants and animals and may include substances which are aggressive to building materials.

Sites previously occupied by plants manufacturing coal-gas, sewage works, chemical works, and scrapyards will be contaminated by the nature of the process involved. In the short term, strong acids or alkalis, toxic organic chemicals, and inflammable or asphyxiating gases may present hazards to human health. In the long term, there may be hazards from the uptake of toxic elements such as cadmium, lead, or nickel from food crops grown in contaminated soil, or through the presence of carcinogens in the soil.

Features of derelict and contaminated land

In the past it was assumed that ruination of the environment in the vicinity of mines and factories was the price which had to be paid for successful industrial activity. It was an attitude which was summed up by the expression 'where there's muck there's money'. Attitudes die hard and attempts to improve the industrial scene are slow to manifest themselves. The dangers of old buildings collapsing, quarry faces crumbling, water-filled pits causing drowning, are only now a matter of public and private concern in Britain. Public awareness of such problems was heightened in 1966 by the disaster at Aberfan when 116 children and 28 adults were engulfed by a flow of colliery waste from a valley-side waste tip of coal shale.

There are less direct consequences of derelict land. Its presence detracts from the image of any neighbourhood so new industry is not attracted to areas of intense dereliction. The visual impact of derelict land engenders disrespect for the surroundings, encouraging vandalism which further degrades the environment.

The main problems (after Goodman 1974) that occur singly, or more usually in combination, on derelict land are:

1. **Site topography.** Much derelict land is associated with holes in the ground or heaps of waste material dumped on the surface. Derelict sites frequently have a complex topography which may require costly earth-moving before redevelopment can take place.
2. **Site stability.** Waste materials tipped on derelict land may be mechanically unstable, especially where they are not freely drained, and may pose a threat to adjacent areas. Often there is only a slight vegetation cover, so rain and wind may wash or blow material away from the site. The action of frost may comminute the material and make it more erodible. Combustion, either spontaneous or as a result of irresponsible human action, results in fumes and unstable surfaces through which it is possible to fall or to be badly burnt. Sediment in the water draining from tip complexes may pollute watercourses. The overall appearance of waste tips is unattractive.
3. **Contamination.** Many derelict sites contain toxic materials which can be injurious if ingested by animals or taken up by plants. Some metallic elements are poisonous, other substances are corrosive or carcinogenic—acids and sulphates may attack and weaken building structures.
4. **Drainage.** Activities such as excavating or tipping may upset the natural drainage of a site and can result in flooding. Where minerals such as coal or salt have been mined, surface subsidence may follow as the ground settles and may upset the natural surface drainage.
5. **Ownership.** Frequently land in urban and industrial areas became fragmented for small businesses. Ownership of these small plots may be disowned, or the records lost. Before redevelopment can take place, the ownership of the land must be consolidated; this may lead to lengthy legal investigations.
6. **Access.** Many derelict sites formerly associated with canal or railway facilities now lack suitable road access.
7. **Ruined buildings.** Frequently there are abandoned, dilapidated, or ruined buildings on derelict land. Some of these have deep, often complex, foundations which may cause problems in redevelopment, even after the clearance of the site to ground level. There may be buried ducts, pipes, and tanks holding toxic liquids (Fig. 1.4).
8. **Unsightliness.** The industries which formerly occupied derelict sites were active when legislation to control pollution was non-existent, or more permissive than today. As a result, contamination by phytotoxic substances may leave a site denuded of vegetative cover. The ruined buildings and gullied tips emphasize present neglect (Fig. 1.5).
9. **Absence of services.** It is expected that all major services will be available on a site to be redeveloped. Usually these facilities are lacking on a derelict site; to bring them in is an additional cost in redevelopment.
10. **Poor housing.** Although not necessarily part of an industrial derelict site, the neighbouring housing may be run-down. Such property, often poorly built and maintained, contributes to the extent of inner city dereliction.

The unattractiveness of derelict land is often enhanced by the presence of active scrap-yards and other unsociable forms of industry. Lack of respect for the area leads the unscrupulous to dump old cars, mattresses, and other domestic rubbish, which then accumulates and is not cleared by official refuse collectors. No one would willingly move into such an area and the ambitious will depart, leaving the elderly or disadvantaged to endure worsening conditions. Falling property prices and a falling revenue from local rates may cause the local administration to give

Fig. 1.4. Ruins of the White Rock Copper Works (in the foreground), Swansea, 1962.

Fig. 1.5. Copper slag heaps and treeless eroded valley-side slopes, Lower Swansea Valley, 1962. Preparations for tree-planting can be seen in the foreground.

such areas low priority for the limited funds available. The vicious circle thus established may be difficult to break.

However, land is a scarce, valuable natural resource, and a policy of *laissez-faire* is unacceptable. Without reclamation, the nuisance of derelict land will persist and new industrial and housing developments will be forced on to fresh sites outside the existing urban areas. If sites could be reclaimed and industrial land recycled, this form of urban encroachment could be halted.

2 Classification of derelict land

The discussion on the definition of derelict land (p. 2) revealed that many words and phrases with imprecise meaning have been used to describe derelict land. The lack of an agreed vocabulary of defined terms has led to inconsistencies in the figures collected by government statistical departments and substantial areas of derelict land in Britain have fallen outside the scope of legislation intended to assist its reclamation.

The character of derelict land varies from one part of the country to another. In South Wales it has been caused mainly by coal-mining or activities directly linked with mining; in North Wales it resulted from slate-quarrying and mineral exploitation. In England many river valleys have been spoilt by gravel-workings; chalk has been extensively quarried in some counties; the extraction of china clay has affected 1700 ha in Cornwall; industrial dereliction from a wide variety of metal-working activities has affected the Midlands; and the chemical industry has affected parts of Cheshire and Lancashire. Many classifications of derelict land were developed for specific localities, and most of these lack the comprehensiveness necessary for national adoption. However, in the last decade, government interest, through the Department of the Environment, has resulted in a more uniform approach.

Many of the published classifications are based on the features of derelict land presented in Chapter 1. Since much dereliction takes the form of a heap or a hole, site relief is often taken as a starting point for classification. Further subdivisions have been based on the character of materials on the site or its capability for sustaining plant growth, the water relations of the site, which have often been upset, resulting in waterlogging or flooding, and the type of industry which previously occupied a site, as this will determine the nature of the ruined buildings and the materials likely to occur there.

Classifications of derelict land have to fulfil two immediate functions—to arrange sites into logical, practical groupings and to act as a map legend. Most classifications attempt to achieve both of these functions, particularly as derelict land has a geographical distribution and it is sensible to present this cartographically. Figure 2.1 is such a map.

Beaver's classification

Beaver (1946) described derelict land in terms of four groups of criteria and categorized sites on maps by a four-digit reference number, where the four digits represented relief, drainage, vegetation, and the composition of the surfaces. Thus (Table 2.1), the code 5312 in Beaver's classification refers to an area

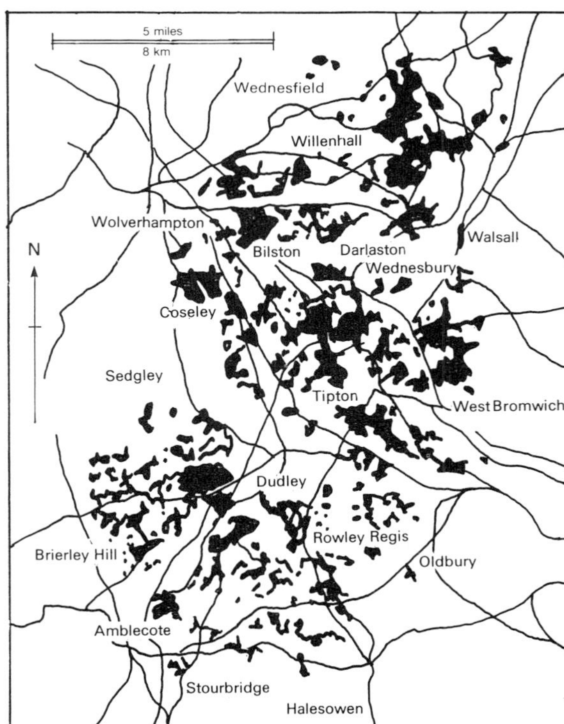

Fig. 2.1. Derelict land in the Black Country (Beaver 1969).

Table 2.1. Beaver's (1946) four-digit classification of derelict land

Relief of surface (first digit)

1. More or less level
2. Gently sloping (i.e. gradient not more than 1 in 20)
3. Steeply sloping (i.e. gradient over 1 in 20)
4. Level, but potholed by subsidence
5. Irregular mounds and hollows (amplitude under 10 ft)
6. As 5, but amplitude over 10 ft
7. Large spoil banks, projecting from surface—conical
8. As 7, but other shapes
9. Marl holes, quarries

Drainage (second digit)

1. Permanently waterlogged
2. Liable to flood.
3. With waterlogged hollows
4. Generally free of surface water

Vegetation (third digit)

1. Bare, little or no vegetation
2. Mainly weeds
3. Weeds with sufficient grass to provide a scanty grazing
4. Grass, weeds, and brambles, or other small bushes
5. Mainly trees
6. Trees

Composition of the surface (fourth digit)

1. Shale
2. Burnt shale
3. Shale mixed with stones and other debris
4. Stones or quarry waste
5. Blast furnace or other slag
6. Rubble, bricks, concrete, etc.
7. Chemical waste
8. Ashes and cinders
9. Town or other domestic refuse

of derelict land with irregular mounds and waterlogged hollows, with little or no vegetation, and a surface layer of burnt shale.

It was recognized that the general appearance of the area surrounding the derelict land was also significant, so one or more of the following letters was added to signify:

A. Houses in bad structural repair.
B. Other houses.
C. Clean factories (not more than one smoking chimney).
D. Factories emitting much smoke.
E. Offensive works or factories, e.g. chemical works.
F. Railway lines.
G. Stream, river or canal.
H. Road.
J. Farmland.
K. Access by main road direct to site.
L. Access by minor road.
M. Access by footpath only.

Beaver suggested that sites might also be classified according to their ease of restoration:

1. Relatively level sites that could be used for housing, industry, playing fields, etc. with little levelling.
2. Sites requiring considerable effort in clearing and levelling.
3. Sites where the work involved would be so great as to cast serious doubts on the feasibility of the reclamation scheme.
4. Great mounds requiring individual attention.

Classification of derelict land in the West Riding of Yorkshire

A survey of derelict land in the West Riding of Yorkshire in 1965–66 at 1:25 000 (Wallwork 1974) adopted a simple legend in which topography was again an important criterion in the classification:

1. **General**
   a. Degenerate, unattractive environment.
   b. Attractive areas and country.
2. **Topography**
   a. Land at natural ground level (but with topsoil removed).
   b. Tipped or filled to higher than natural ground level.
   c. Excavated land.
3. **Potential dereliction**
   a. Land where mineral waste-tipping is permitted under the consent of the planning authorities.
   b. Land where mineral extraction continues under the consent of the planning authorities.
   c. Land liable to flood.
4. **Nature of the waste**
   a. Mineral extraction waste.
   b. Colliery waste.
   c. Power station waste.
   d. Domestic waste and hard core.
   e. Industrial waste.

This classification did not specify the nature of the waste in detail, which limited its usefulness.

Oxenham's classification

Oxenham's (1966) classification also began with site

relief. He divided sites into those upon which materials had been deposited to create a mound, or land from which materials had been extracted to leave a hole. The categories (Table 2.2) lack consistency and may be difficult to apply in a standard manner throughout the country.

Later, Oxenham (1967) suggested that, since modern earth-moving machinery could landscape even the largest heaps in a relatively short period of time, the form of the derelict landscape no longer limits restoration. The limiting factor now became the cost of the earth-moving operation.

Table 2.2. Oxenham's classification

A. Mounds and spoil heaps	B. Pits and excavations
1. Colliery spoil heaps a. High conical b. Ridge c. Low flat	1. Dry excavations
2. Quarry spoil heaps a. Oil shale heaps b. Stone-quarry waste heaps c. Clay and sand heaps	2. Flooded excavations
3. Cement, glass, pottery, and chemical wastes	3. Quarries
4. Slag heaps 5. Ash and clinker heaps 6. Hill and dale formations from opencast mining	4. Abandoned industrial sites

Surveys by the Department of the Environment

There were annual surveys of derelict land in England between 1964 and 1971 and in Wales between 1964 and 1969. These surveys recorded derelict land included in the official definition (p. 2), but excluded land in the categories detailed on pp. 2 and 3. They identified spoil heaps, excavations, and 'other forms of dereliction'. They estimated the proportion of the derelict land in each category which justified restoration. The data were compiled for each administrative county and can be obtained from the Department of the Environment (DoE) or the Welsh Office for England or Wales, respectively.

As a result of experience gained from the pre-1971 surveys, the Department of the Environment proposed a new survey of derelict land in 1973. Changes in the structure of local government delayed the survey until 1974 and the results were published in three volumes, one of which summarizes the survey (DoE 1975). Each administrative area of England was asked to record the derelict land within its boundaries on 1 April 1974 under three major headings: inherited derelict land, abandoned mineral workings, and refuse tipping (Tables 2.3 and 2.4). The questionnaire also solicited information about the extent of each currently affected by the extraction of various minerals and attempted to ascertain the increase in derelict land between 1 January 1972 and 31 March 1974, as well as the total restored during that period, and the extent of restoration anticipated between 1 April 1974 and 31 March 1976.

There was a further survey of England in 1982. It excluded neglected woodland, farmland, natural unused land, and damaged land which has blended into the landscape so that it is no longer a problem, and collected no information on sites being used for current tipping and those which await development. The results were presented as tables of the amount of derelict land classified as spoil heaps, excavations and pits, military dereliction, derelict railway land, and 'other forms of dereliction' (DoE 1984). The area in each county justifying restoration is given (Table 2.5).

Table 2.3. Derelict and despoiled land in England as of 1 April 1974—inherited dereliction and refuse-tipping

Inherited dereliction and refuse-tipping	Area (ha)
Total area of inherited dereliction	43 272
Of which, areas justifying restoration are	
Spoil heaps	9 084
Excavations and pits	6 596
Military and other Service dereliction	3 145
Abandoned British Rail land	6 412
Other forms of dereliction	7 831
Total justifying restoration	33 068
Total increase in area of inherited dereliction since previous survey (31 December 1971)	10 309
Total area of abandoned land damaged by mineral workings now ceased but where restoration is unfulfilled	2 307
Land used for public refuse and private and commercial tipping of non-mineral waste	
Total area currently in use	11 399
Of which, area not covered by restoration conditions	2 959

Table 2.4. Derelict and despoiled land in England as at 1 April 1974—mineral working

	Land with permission for mineral working (ha)	Amount of land currently affected by excavations, tips, etc. (ha)
Analysis of total area with permission for mineral workings		
Area covered by restoration conditions	73 490	36 196
Area not covered by restoration conditions	20 912	12 880
Total area with permission for mineral workings	94 402	49 076
Analysis by minerals		
Chalk	3 696	2 132
China clay	4 956	1 749
Clay and shale*	11 875	5 417
Coal	11 164	8 925
Gypsum/anhydrite	1 009	436
Igneous rock	2 771	1 347
Ironstone	6 051	1 220
Limestone	10 698	5 451
Sand and gravel	33 401	18 171
Sandstone	2 269	1 146
Silica and moulding sand	3 899	1 985
Slate	169	111
Vein minerals†	1 326	621
Other minerals	1 148	395

*Including fireclay, ball clay, and potters' clay.
†Including tin, copper, lead, silver, zinc, haematite, iron ore, barytes, calcspar, fluorspar, and witherite.
(This table is Crown Copyright and reproduced here by permission of the Department of the Environment, from DoE 1975.)

Table 2.5. Derelict and despoiled land in England as at 1 April 1982 (in hectares)

	Area (ha)
Total area of inherited dereliction	45 683
Of which, area justifying restoration	
Spoil heaps	8 300
Excavations and pits	6 402
Military and other Service dereliction	2 452
Abandoned British Rail land	6 015
Other forms of dereliction	11 109
Total justifying restoration	34 278

(This table is Crown Copyright and reproduced here by permission of the Department of the Environment, from DoE 1975.)

The report also attempted to indicate the proportions of urban/rural derelict land, and also the ownership of the land, under the headings (i) local authorities and (ii) other public bodies, with the assumption that the balance is privately owned. A separate survey of mineral sites for England and Wales was recently published (DoE 1984; Welsh Office 1985). Table 2.5 is directly comparable with the first part of Table 2.3.

Downing's check-list

Downing (1977) presented a check-list for individual sites to be used in regional surveys of derelict land (Table 2.6). The categories conform generally with the DoE surveys and are defined by general and specific information on the cause of dereliction, physical form, and vegetation cover of the site. The check-list can also be used for preliminary site assessments, and can provide a classification for interpretations of aerial photographs of derelict sites.

The classification of Collins and Bush

The classification system (Table 2.7) proposed by Collins and Bush (1969*a*, *b*) resulted from studies on

Table 2.6. Downing's (1977) check-list

Site name	Location and map reference				
Cause of despoilation (general)	**Cause of despoilation (particular)**	**Physical form (general)**	**Physical form (particular)**	**Specify volume where appropriate**	**Vegetational cover**
1 Mineral working ☐	1 a Chalk ☐	Each of categories in Column 2 gives rise to one or more of the following generalized forms:	A 1 Ridge ☐		1 None ☐
2 Tipping ☐	b China clay ☐		2 Cone ☐		2 Sparse ☐
3 Industry ☐	c Clay and shale ☐		3 Other shape ☐		3 Ephemerals ☐
4 Transport ☐	d Coal ☐	A Heap ☐	B 1 Flat ☐		4 Herbs only ☐
5 Services (military etc.) ☐	e Gypsum/ anhydrite ☐	B Ground level (includes low heaps and shallow excavations not more than 3 m (10 ft) from original ground level) ☐	2 Ground tainted by chemical wastes etc. ☐		5 Herbs and shrubs ☐
6 Others ☐	f Igneous rock ☐	C Excavation ☐	3 Ground affected by derelict buildings, foundations, or other derelict artefacts ☐		6 Herbs and trees ☐
	g Ironstone ☐		4 Hills or hollows ☐		7 Shrubs and trees only ☐
	h Limestone ☐		Rectangular ☐		
	j Sand and gravel ☐		C 1 Other shape (wet or dry) ☐		
	k Sandstone ☐				
	l Silica and moulding sands ☐				
	m Slate ☐				
	n Vein mineral ☐				
	o Other minerals ☐				
	2 a Public waste ☐				
	b Commercial waste ☐				
	c Others ☐				
	3 a Brickworks ☐				
	b Chemical works ☐				
	c Gasworks ☐				
	d Iron and steel ☐				
	e Power station ☐				
	f Sewage works ☐				
	g Others ☐				
	4 Airfields (private) ☐				
	Canals ☐				
	Railways (BR) ☐				
	Roads ☐				
	Others ☐				
	5 Airfields (military) ☐				
	Camps ☐				
	Defence establishments ☐				
	Others ☐				
	6 Others (specify) ☐				

Note: Separate forms would be used for individual sites or identifiable parts of sites.
(This table is from Downing 1977.)

the usefulness of aerial photography in certain aspects of planning. It is comprehensive. They define derelict land as '. . . land which has been so damaged by extractive or other industrial processes that it gives offence to the eye and is likely to remain so until subjected to special forms of treatment'. Their classification is applicable anywhere in Britain, and it can be related to the categories employed in the DoE (1974) survey. They describe a site by a symbol of up to six characters, and this can be used to annotate maps and air photographs or to record practical assessments of derelict sites. A capital letter to indicate the general topography is followed by a figure for the activity associated with the site, and a Roman numeral for the specific cause of dereliction. Available filling materials which might be of use in restoration are listed after an oblique line. Using this classification, A1ia signifies a disused coal tip and

Table 2.7. The classification of Collins and Bush

Topography

A. Dereliction above ground level—heaps
B. Dereliction below the natural ground level—holes
C. Dereliction at ground level
D. Installations

Associated activity—general	Associated activity—specific
1. Mineral working	i. Coal
	ii. Brick clay
	iii. Lead
	iv. Ironstone
	v. Limestone
	vi. Chalk
	vii. Sand and gravel
	viii. China clay
	ix. Tin
	x. Slate
	xi. Others
2. Refuse	i. Household waste
	ii. Scrap
	iii. Cars
	iv. Others
3. Industrial workings	i. Brickworks
	ii. Chemical works
	iii. Gas works
	iv. Iron and steel works
	v. Power stations
	vi. Sewage works
	vii Others
4. Transportation	i. Airfields
	ii. Canals
	iii. Railways
	iv. Roads
	v. Others
5. Others	i. Others

Pictorial description

a. Tip
b. Dump
c. Dry pit
d. Wet pit
e. Cleared land but still spoiled or degraded
f. Degraded land associated with the site
g. Degraded land peripheral to the site
h. Opencast workings not yet pits
j. Sludge

Filling materials—general

Refers to the nature of the material used to fill an excavation. The categories are described by the numbers 1 to 5 listed in 'Associated activity—general' above.

Filling materials—specific

The categories are described by the Roman numerals listed in 'Associated activity—specific' above.

B1vii/3v a sand and gravel pit filled with pulverized fuel ash.

A site still in active use, although otherwise derelict, such as a colliery tip, is indicated by a suffix Z.

The Second Land Utilisation Survey

The Second Land Utilisation Survey (Coleman and Shaw 1980) also mapped categories of derelict and transitional land. The legend of the maps makes the following subdivisions, based on the previous use of the land. Active demolition is shown by a further symbol and areas of active tipping are indicated by a red stipple with a letter symbol to indicate what is being tipped.

Categories of derelict land

Nurseries	Tips
Agricultural buildings	Roads
Houses	Vehicle parks
Chalets	Railways
Flats or joint residences	Ports
Businesses	Airfields
Institutions	Water or drained areas
Factories	Unidentified buildings
Utilities	Unidentified land
Quarries	

Categories of active tipping

C. Coal shale tips
D. Tips for dredging
F. Factory spoil tips
M. Mine tips other than coal
N. Tips for noxious substances
Q. Quarry spoil heaps, i.e. tips from any surface mineral extraction
R. Refuse tips
T. Temporary tips, i.e. part of a reclamation process
U. Spoil tips for public utilities
X. Unofficial refuse tipping

Wallwork's classification

Wallwork (1974) suggested that a comprehensive survey of derelict land should include information on the source of dereliction, topography, amplitude of dereliction, composition of materials, vegetation cover, and potential for reclamation. If it could be objectively measured, he would also record ugliness. Wallwork has been particularly interested in the Cheshire district where he classified derelict sites in terms of letter and number symbols which can be annotated on aerial photographs or maps (Fig. 2.2). He assembled the symbols in the form of a fraction

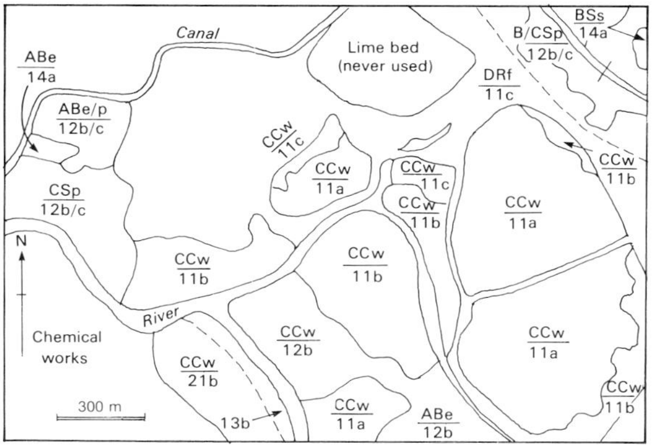

Fig. 2.2. Mapping of derelict land near Northwich (Reproduced by permission of the publishers, David & Charles, from Wallwork 1974).

with symbols for the type of dereliction, kind of industry, and form of site in the numerator and for the amplitude, relief, and vegetation cover in the denominator, e.g.

$$\frac{\text{CCw}}{\text{lla}} = \frac{\text{Mineral-based industry: chemicals: waste}}{\text{Less than 5 m amplitude: level: unvegetated}}.$$

His four main categories are:

A. Surface mineral extraction.
B. Deep mining.
C. Mineral-based industry.
D. Transport.

They are subdivided according to the type of mineral or industry or transport involved in the previous site use, with a further subdivision based on the presence of excavations, wastes, subsidence, or buildings. The amplitude of topography, surface configuration, and vegetation are specified according to:

Amplitude
1. Under 5 m.
2. 6–30 m.
3. More than 30 m.

Surface configuration
1. Level or gently sloping.
2. Pitted or undulating.
3. Gullied or steeply sloping.
4. Waterlogged or flooded.

Vegetation
a. Absent or no soil.
b. Rough grassland.
c. Scrub or sparse woodland.
d. Reeds.

Haines's classification

Since then emphasis has changed and the problems of land contamination have assumed a greater significance. Contaminated land is often derelict too, but the two labels are not synonymous. Classification according to the degree of contamination is not easy as there is a wide range of possible contaminants of different degrees of toxicity. However, guidelines have been issued by the Interdepartmental Committee on the Redevelopment of Contaminated Land (ICRCL 1983), and are referred to in Chapter 4 (p. 37). Haines (1981) attempted to classify categories of contamination according to the site use, on the assumption that site use, pollutants, and land contamination are linked:

1. Definitely contaminated.
2. Probably contaminated.
3. Potentially contaminated.
4. Pre-1931 housing where lead paint might be exposed.
5. Mixed land use with many small potentially contaminated sites.
6. Definitely uncontaminated.
7. Unknown.
8. Previously contaminated sites that are now in 'sensitive' uses.

Discussion

From this review of classifications it may be concluded that surveys of derelict land can be presented on maps at different scales. Certain national and regional statistics may be depicted on small-scale maps with broad classifications (e.g. Figs. 1.1–1.3), and local and district information may be assembled using medium scales. However, plans for restoration are usually drawn up for individual sites, and require that accurate details of quantities of material and hazards be mapped on a large-scale map. The systems of classification reviewed in this chapter cannot achieve the detail or accuracy necessary. Thus Chapter 3 considers the patterns of contamination associated with several major categories of industry, Chapter 4 discusses the components of derelict and contaminated land, and Chapter 5 outlines methods for surveying individual sites in detail.

3 Derelict land and patterns of contamination

When sites of mineral extraction are exhausted or industrial plant becomes obsolete and is abandoned, land becomes derelict. Waste materials and contaminants may have been dumped on site—their nature and disposition will influence the choice of procedures for reclamation. Many different combinations of features and contaminants are possible so each site must be treated individually. Whilst derelict land can result from any form of land use, not all uses result in land contamination. This chapter discusses the patterns of contamination and dereliction likely to be associated with different kinds of derelict industrial site. As a framework it uses the standard Industrial Classification of the Central Statistical Office (1980) which divides industrial and other activities into 10 broad categories (divisions):

0. Agriculture, forestry, and fishing.
1. Energy and water supply industries.
2. Extraction of metals, manufacture of metals, mineral products, and chemicals.
3. Metal goods, engineering, and vehicle industries.
4. Other manufacturing industries.
5. Demolition and construction.
6. Distribution, hotels and catering, repairs.
7. Transport and communications.
8. Banking, finance, insurance, and business services.
9. Other services.

The World Health Organization has published a parallel classification of industrial activity (World Health Organization 1982), which lists industries likely to contaminate the environment, together with a means for estimating the amounts of contamination released in air, water, and land pollution. The Interdepartmental Committee on the Redevelopment of Contaminated Land (ICRCL 1983) lists the kinds of site on which hazardous concentrations of contaminants are likely to be found. This list includes:

Landfill and other waste-disposal sites;
Gasworks and coal carbonization plant and ancillary by-product works;
Sewage works and sewage farms;
Scrap-yards;
Railway land, especially large sidings and depots;
Oil refineries, petroleum storage and distribution sites;
Metal mines, smelters, foundries, steel, and metal finishing works;
Chemical works, including paint, plastics, and pesticide works;
Munitions production and testing sites;
Asbestos works;
Tanneries;
Paper and printing works;
Industries using or making wood preservatives.

Haines (1981) lists many of the pollutants emitted from a range of industries, and Patterson (1979) lists pollutants from metalliferous industries.

Agriculture, forestry, and fishing

Agriculture and forestry normally maintain the soil and land in a productive state, and so do not produce dereliction except through bad or unsuitable agricultural practices, e.g. severely gullied land and old farm buildings unsuitable for modern farming methods which have been allowed to become derelict. Some soils have become contaminated by prolonged and excessive use of pesticides in vineyards and orchards, but as a general rule agriculture and forestry are not creating derelict land. Any dereliction attributable to fisheries is likely to be associated with docklands.

EXPECTED PROBLEMS: acidity, buildings, pesticides, soil erosion. (See also *Docks*, p. 33.)

Energy and water supply industries

No great problems of dereliction have resulted from the water supply industry. However, many examples of dereliction have arisen from the extraction of coal from deep mines and opencast workings and its subsequent processing, including coke and gas manufacture; the extraction of mineral oil and natural gas; the generation of electricity; and the use of nuclear energy.

Coal—deep mine sites

Since coal was first used in the twelfth and thirteenth centuries, methods of extraction and the siting of the mines have changed. Initially coal was obtained from bell pits, excavated on the outcrop of the coal seam. A hole was dug into the coal and extended laterally until the sides were in danger of falling inwards. Such workings may be identified by the hummocky ground surface and they can still be seen in old mining areas.

During the nineteenth century it became necessary to excavate shafts down to deeper coal seams and the mining industry gradually changed its location to follow the dip of the coal-bearing rocks. The exploitation of 'concealed' coalfields began during the twentieth century and this is now the major field of production (Fig. 3.1). Many mines exploiting the deeper seams of the 'exposed' coalfields ceased production in the 1950s, contributing greatly to the stock of derelict land.

When abandoned, the sites of deep mines often contained ruined buildings and large waste tips (Fig. 3.2). For safety, the actual shaft was usually filled and capped with a layer of concrete. By now much of the waste has been used as fill for road foundations and many of the tips which remain are often landscaped so that their harsh lines blend into the surroundings. Much of the coal was moved by rail so there will be areas of railway sidings even though the branch line which served the colliery has long since ceased to operate. Away from the immediate pit head, there may be subsidence, causing damage to buildings and flooding. Drainage from mines can be strongly acidic and ferruginous through the oxidation of pyrites, and may pollute watercourses and kill aquatic life.

EXPECTED PROBLEMS: acidity, buildings, colliery waste, combustible materials, dust, erosion and lack of soil, subsidence, and flooding. (See also *Transport and communications*, p. 32.)

Coal—opencast sites

Opencast (or strip) mining involves the complete removal of the overlying rock down to the coal. In the older areas of the coalfields, the coal in deep mines was worked by the 'pillar and stall' method which left almost half the coal to support the roof of the underground workings. Where these old workings are at shallow depth, opencast mining can be used to remove the remaining coal. At the same time any surface dereliction resulting from previous activities can be buried during restoration. The depth of these opencast excavations is related to the type of equipment used and the economic ratio of overburden to coal extracted. When opencast coal mining began,

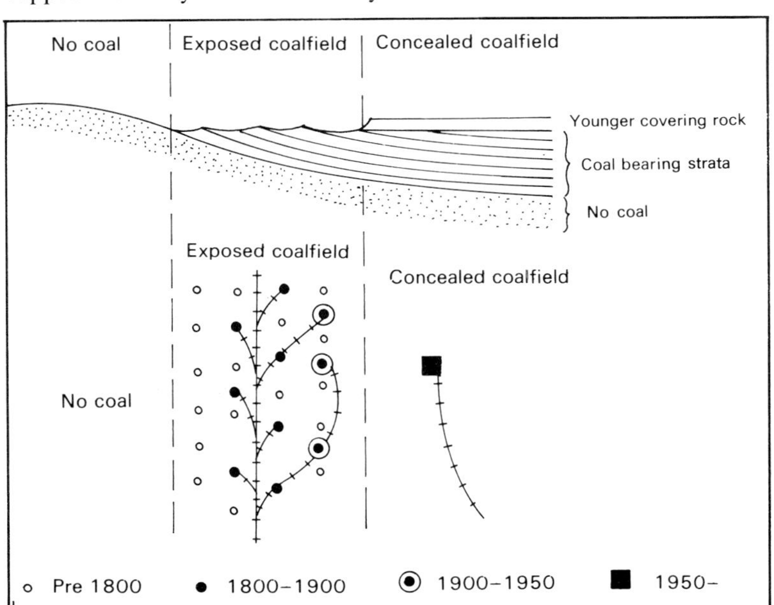

Fig. 3.1. Spread of mining across a coalfield.

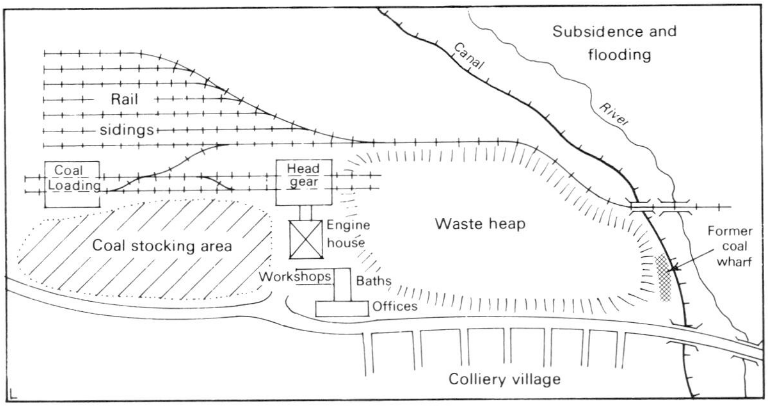

Fig. 3.2. Diagrammatic plan of a coal-mine.

Fig. 3.3. Coal extraction by opencast methods, Derbyshire.

workings were only about 15 m deep and the working ratio of overburden to coal was approximately 4:1. In the 1950s it became possible to work to depths of 100 m with a ratio of 20:1 and since the 1960s depths of 150–230 m have been worked with a ratio of 40:1. Where anthracite is mined the ratio may rise to near 70:1 (Fig. 3.3).

Compared with underground mining, opencast mining is less expensive, safer, and the working conditions are more pleasant. In the short term it devastates the landscape completely, but it is possible to replace the overburden and to recontour the land. Opencast mining was introduced into Britain as an emergency measure in 1942. From the outset the government insisted on restoration of the sites worked so opencast mining has not created long-term dereliction.* In the USA, where opencast mining is referred to as 'strip' mining, many areas have not been reclaimed. In 1974, the US Department of the Interior stated that over one million hectares required treatment. The total has probably increased since

*Open Cast Coal Act, 1958 (amended to Coal Mining Industries Act, 1975), and recently Town and Country Planning Acts (summarized in DoE Circular 3/84 and Welsh Office Circular 13/84). All restrict operations and lay down conditions for restoration after open cast mining.

then and it was expected that a further 172 000 ha would be mined between 1980 and 1985. This large extent of derelict land developed because the regulations have been ineffectual or weakly enforced in many states (Bradshaw and Chadwick 1980).

In Czechoslovakia, coal-mining had devastated 25 000 ha by 1969 and this was expected to amount to 41 547 ha by 1980. Reclamation of 21 400 ha was planned for the period up to 1980 (Jonas 1973). Between 1880 and 1950, lignite mining affected 47 636 ha in East Germany; only 46.8 per cent had been reclaimed in 1950 and each year produces another 1200 ha of derelict land. A further 300 000 ha of mineable lignite remain (Knabe 1973). In West Germany brown coal deposits cover 3000 km^2 between Cologne, Aachen, and Dusseldorf with a reserve of 60 billion tonnes. Up to 1968, 15 000 ha had been disturbed by opencast mining.

After the presence of coal reserves has been proven, a cut is opened down to the coal at one end of the site and the coal removed by mechanical shovel and lorries. Mining then proceeds in stages across the site. The first stage is the removal of topsoil and subsoil by scrapers. The overburden is blasted where necessary and then moved by dragline across the cut on to the strip from which the coal has been removed (Fig. 3.4). Topsoil and subsoil are stored separately in large banks around the periphery of the site where they are used to screen operations from neighbouring property (see p. 96). Once the process is under way a well-organized site will transfer the successive strips of topsoil and subsoil removed directly on to a worked area which has been levelled to receive it.

Opencast coal-mining has many advantages despite the temporary devastation. It enables the whole site to be cleared, usually in a period of five to 10 years and the scars of former dereliction to be removed completely. Old tip heaps and any other unsightly or toxic material can be buried deeply and a completely new landscape created. Problems arise when the operation creates high cliffs or steep slopes which subsequently are difficult to reclaim. This has happened in the Appalachian coalfield of the USA, where contour strip mining has been used (Fig. 3.5).

EXPECTED PROBLEMS: acidity, colliery waste, combustible material, dust, quarry faces, soil erosion and lack of soil, and subsidence.

Thermal power stations

Although the consumption of electricity has risen steadily since the Second World War until the present economic recession, rationalization of the industry has phased out some of the smaller thermal power stations. Consequently, a number of sites have become available for redevelopment, often in locations very suitable for new industrial activity.

In 1952 there were 274 small power stations in England and Wales; by 1970 there were 187, of which 136 were coal-burning; currently there are 108 thermal power stations. In 1970, the 136 coal-burning stations were using 67 million tons per year and produced 12.1 million tons of ash and clinker (Gutt *et al.* 1974). Before 1930, all stations burnt lump coal which left coarse ash and clinker. Clinker consists of glassy masses with inclusions of refractory materials and nodules of unburnt coal. It may be used for aggregate or fill road works. From 1930 there was a progressive change to pulverized coal and more than half of the present stations, including some of the largest and most modern, are fired by pulverized fuel. In these power stations, coal is finely ground and burn in a current of air, the resulting pulverized fuel ash (pfa) is composed largely of small glassy spherules.

A former power station site may include the remains of buildings; tall chimneys or heavy machinery will have required substantial foundations which may remain below ground level. There may be former railway sidings and coal dumping areas, with cooling towers and underground pipework (Fig. 3.6).

EXPECTED PROBLEMS: asbestos, buildings, combustible materials, lagoons, polychlorinated biphenyls, pulverised fuel ash. (See also *Transport and communications*, p. 32.)

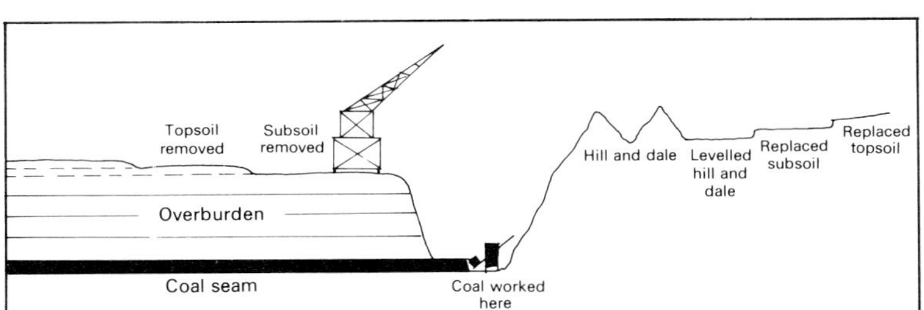

Fig. 3.4. Strip mining by drag-line—cross-section.

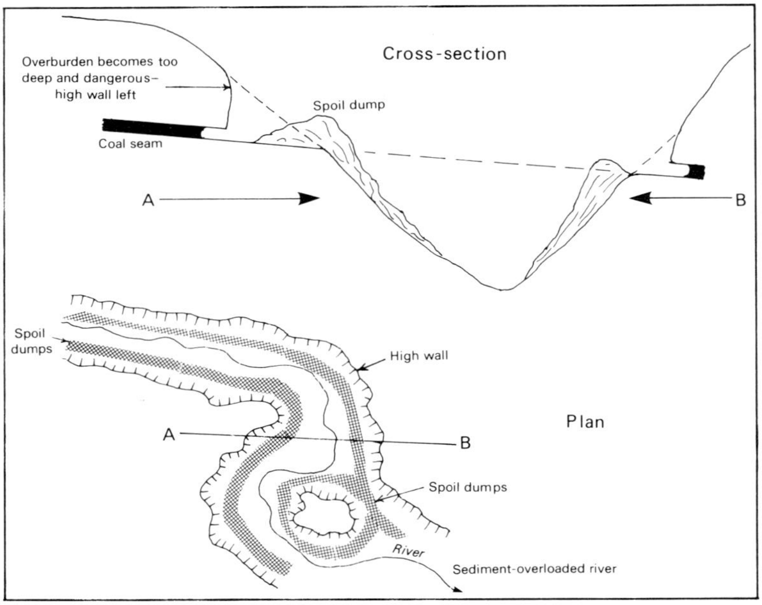

Fig. 3.5. Contour strip mining.

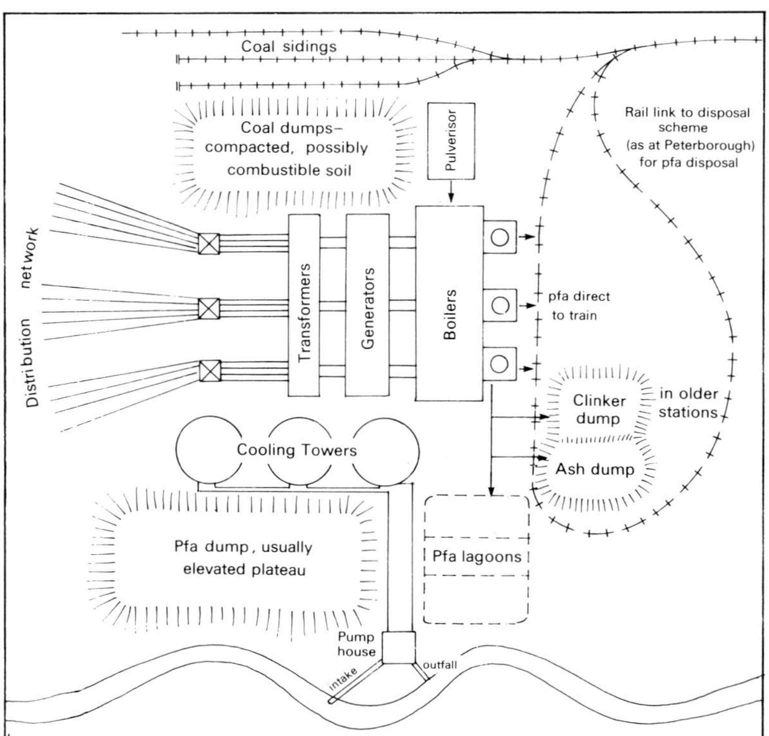

Fig. 3.6. Diagrammatic plan of electric generating station.

Carbonization plants

Carbonization of coal produces four main products: gas, coke, tar and ammoniacal liquor and production of any one of these can be maximized by adjustment of the process.

Many small gasworks became obsolete when the gas industry was rationalized and many more town gas plants became redundant upon the change to natural gas. There are still a limited number of coke and tar works associated with steel works, but contraction of the iron and steel industry has left many of their sites available for redevelopment too (Dean and Goalby 1980; Wilson and Stevens 1981).

The redundant plant may have been demolished and the site levelled; this may have spread pollutants over the whole site. Buried gasholder wells, tanks, and pipework may be troublesome when future development takes place. Below-ground structures may hold toxic materials or act as pathways for the migration of any remaining liquid or gaseous contaminants.

The solid contaminants include tars and the ferric oxides used to remove hydrogen sulphide, cyanides, and ammonia from the gas. Liquid contaminants may include toxic or even carcinogenic materials which may seep into the ground to contaminate aquifers (Young and Barber 1980), so they should have been removed from the site when it was dismantled. Buried tars are likely to include benzene, toluene, xylene, ethyl benzene, styrene, phenol, cresols, xylenols, polyhydric phenols, naphtha, naphthalenes, acenaphthalene, fluorine, diphenol oxide, anthracene, phenanthrene, carbozole, tar bases, and medium soft pitch (ICRCL 1979*a*).

Site contaminants may have been pushed into old gasholder wells and capped with a layer of clay or concrete to stop influx of water and to prevent overflow and dispersal of the contaminants. Some of them may attack and weaken concrete structures. Also, former gasworks may have been stripped of asbestos insulation and the discarded asbestos buried (Fig. 3.7).

The site may also contain combustible materials such as unburnt coal, dust, coke, and partially carbonized coal. The sulphur content of the spent oxides may be sufficient for them to ignite. Once alight, toxic fumes would also be released from tars and any lighter fractions from the carbonization process which remain in the debris. A gasworks smell may pervade the whole site.

EXPECTED PROBLEMS: acidity, asbestos, buildings, combustible materials, cyanide compounds, phenols, rubble, solvents, spent oxides, sulphates.

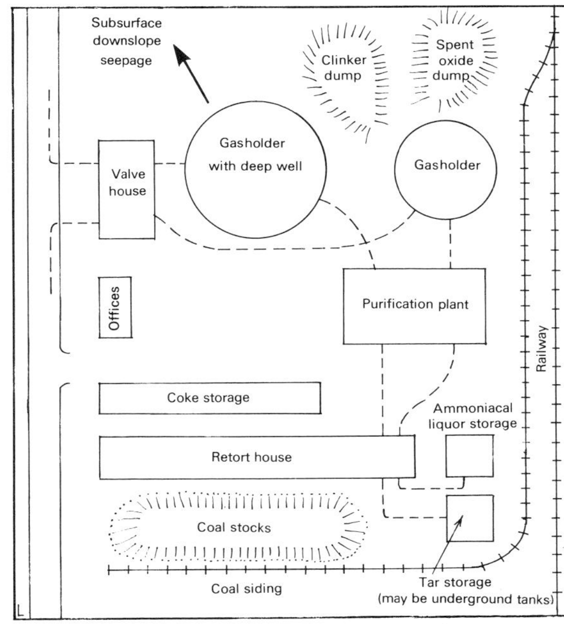

Fig. 3.7. Diagrammatic plan of a town gasworks.

Petroleum and oil extraction and refining

Oilfield sites will probably be contaminated with crude oil, drilling muds, and brines. Accidental spills and blow-outs may occur less frequently now than in the past but many oilfield sites must be saturated with oil from spillages. Away from the oilfields, oil and petroleum products may be spilt anywhere they are handled, stored, or refined. Accidents and spillage also occur from road or rail tankers.

Oil refineries have grown in size steadily with the demand for petroleum products. Expansion of older, smaller refineries has extended outwards from the original sites, which were isolated from other competing land uses, so few sites have become derelict. A refinery is designed to deal with inflammable liquids, often under pressure and elevated temperatures. It consists of storage tanks for the incoming crude oil, interlinked with pipes to transfer oil from one tank to another or to the distillation or cracking plant. Fractionation of the crude oil produces many products which are purified and then stored on a tank farm.

All these operations take place under strictly controlled conditions and accidents and spillages are few. Some inevitably do occur and land has been saturated with petroleum products or their burnt remains. It is normal practice to surround tanks with banks to prevent contamination spreading should an accident occur.

As well as oily and toxic sludges, oil refineries may produce resins, sulphuric acid, and purification muds. Some of these are incinerated or consigned to landfill, but some may remain on derelict sites. Liquid wastes may contain high levels of organic carbon with a high biological oxygen demand and include ammonia, sulphides, phenols, and hydrocarbons. The persistence of these in underground tanks and pipes must be suspected and the danger of combustible material borne in mind. Oil-soaked ground may give off methane if poorly drained, and various chemical additives, such as tetraethyl lead, may also be present. EXPECTED PROBLEMS: acidity, asbestos, buildings, coal tars, combustible materials, lead compounds, oils, phenols, rubble, solvents, sulphates.

Nuclear energy

Initially, nuclear power appeared to offer unlimited supplies of energy, but controlling the release of energy from a nuclear reaction has not been as easily accomplished as was hoped. Progress in the design and construction of nuclear power plants has been slow and the building of nuclear power stations has provided considerable public controversy and anxiety.

The functional centre of a nuclear power plant is the core where a controlled fission of ^{235}U takes place and releases heat. This heat is carried away from the core by gas (a gas-cooled reactor) or water. The water is contained in a primary cooling loop and may be under pressure (pressure-water reactor) or in earlier designs the water was allowed to boil (a boiling-water reactor). In a boiling-water reactor the steam generated is used to drive turbines, condensed, and returned. In a pressure-water reactor, the water passes through a heat exchanger where its heat is passed to a secondary loop which generates steam to drive turbines. Both forms require large quantities of coolant water, so nuclear plants are located alongside major rivers or the coast. Fast breeder reactors use plutonium, which is potentially more dangerous.

Radioactive contamination of the environment will occur if fission products get into the coolant water, or by the action of neutrons striking substances in solution to generate additional radioactive elements in the water itself. In some cases corrosion in the heat exchanger loops has allowed leakages from the core to the outside. Accidents have occurred during the operation of nuclear power stations in the United States and the Soviet Union. At Three Mile Island, Pennsylvania, radioactivity was contained, but at Chernobyl in the Ukraine, contamination of the area surrounding the power station necessitated an evacuation of local residents. Effects from the fallout of the Chernobyl accident were recorded across northern Europe where crops and livestock were contaminated.

As the uranium in the reactor core decays, the fuel rods become used up, so that after about 10 years they need to be reprocessed. In Britain, this operation is carried out at Sellafield (Windscale) in Cumbria, where there have been low-level radioactive discharges into the Irish Sea and occasional accidents have resulted in high levels (100 times background) of radioactivity being recorded on the beaches of Cumbria. Except for the recorded leakages and permitted discharges, the statistics of derelict land include no categories of land contamination from radioactivity. However, the first generation of nuclear power stations will eventually become obsolete and redundant dockyards dealing with nuclear submarines must also be suspect areas. There are also lesser problems with low-level radioactive materials in industrial waste on landfill sites.

On present policy radioactive wastes are being stored. As they age, they gradually become less dangerous. Elements with short half-lives, measured in days or weeks, do not pose too many difficulties. Elements with longer half-lives are removed from the environment and encapsulated in cement in 55-gallon drums. Currently, there is controversy over where these wastes should be stored.

EXPECTED PROBLEMS: buildings, radioactive substances, rubble. (See also *Thermal power stations*, p. 17.)

Extraction of metals and ores; manufacture of metal, mineral products, and chemicals

The use of metals can be traced back to prehistory. Against the obvious advantages which they bestow must be set the fact that their production has been indirectly responsible for widespread dereliction. Dereliction was thought to be the price paid for obtaining ores and minerals from the earth. In the last 50–60 years it has been realized that machinery which rips the earth apart can also be used in its reclamation and an increasingly vocal environmental lobby has forced governments to allocate funds to rectify past dereliction and to legislate against future devastation.

Dereliction may be caused by the abandonment of industrial activity: (1) where the ore is mined and concentrated; (2) where the ores are smelted; and (3) where there are secondary industrial processes. The first and second of these are considered in this section.

Originally, ores were obtained from rich mineralized zones by means of sloping adits or vertical shafts. Deep mines are still used for many minerals and may require sophisticated machinery to cut and raise the ore. Exhaustion of the richer ores has caused miners to accept ores with progressively lower metallic contents, thus increasing the scale of the operation considerably.

Deep mines

A deep mine consists of vertical or inclined shafts. Local geological conditions dictate the location of the shaft and disposition of the underground galleries from which the minerals are extracted. A deep mine for minerals may be similar to the mines discussed in *Coal–deep mine sites*, p. 15. At the entrance of the shaft, there may be headgear for hauling ore to the surface, an engine house to provide power for haulage, and possibly a stocking area for coal to fuel the engine. These are characteristic of most nineteenth-century mines. However, the most obvious feature of a derelict mine is the waste heap, usually near the shaft entrance (Fig. 3.8).

In most mining operations, the ore is concentrated at the mine site before despatch. This process involves grinding the ore to silt-sized particles and then separating the mineral from the country rock by flotation. This produces a slurry of water and reject material known as tailings which is normally led into lagoons so the rock or gangue material can sediment to the bottom and leave the water available for reuse. Eventually, the lagoons become filled with sediment and are left to dry out. The drying process can take some time and the lagoons can be dangerous if man or machinery ventures on to their thin crusts. Mines from the nineteenth century will probably have railway sidings, and earlier ventures probably a canal wharf. In many cases a mineral mine site will occupy a smaller area than a coal mine site.

Under this category we are concerned with mines for copper, lead, zinc, nickel, silver, as well as the higher-grade iron ores such as haematite.

EXPECTED PROBLEMS: acidity, arsenic, cadmium, chromium, copper, dust, flooded workings, lagoons, lead, nickel, rock waste, rubble, soil erosion, zinc, (See also *Railways*, p. 32.)

Quarries

Where a mineral deposit can be exploited by surface working, a quarry is opened. The nature of the ensuing waste will depend on the rock type. When extracting high-quality building stone and slate, much material is rejected, whereas when stone is crushed for aggregate, or where limestone or chalk are quarried for lime, there will be little waste, as virtually all the stone is usable. Sand and gravel quarries and clay pits for brick and tile manufacture normally do not produce large quantities of waste material and the derelict site is an empty or water-filled hole. Many local authorities utilize old quarries for dumping domestic rubbish. Once used for this purpose, the sites must be classed as landfill (p. 33).

Before this century, quarries were usually small, local operations employing few men and having little impact on the environment. The quarries extended slowly and merged into the landscape as natural

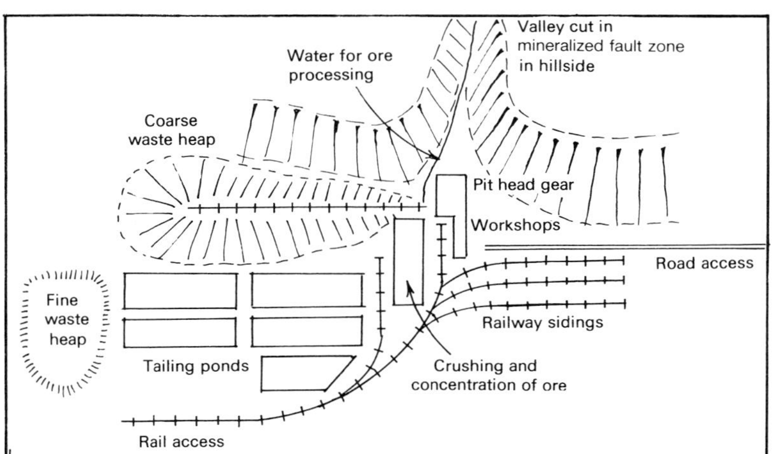

Fig. 3.8. Diagrammatic plan of a mine (other than coal).

regrowth of vegetation took place over the worked areas. As the demand for stone has risen substantially, production has been concentrated at a few large quarries which have had a significant and damaging impact, often in areas of scenic beauty.

The problems associated with derelict quarries include harsh outlines with precipitate faces in some quarries and large quantities of rejected stone in others. Usually there are few buildings associated with the quarrying of stone, but substantial buildings may be present at slate quarries as at Llanberis (now a museum). Kilns and associated buildings may be present where lime or cement is made from quarried limestone or chalk. Quarry floors are normally highly compacted by the machinery and heavy loads carried across them, and there may be accompanying railway sidings (Fig. 3.9).

Sand and gravel quarries, as well as clay pits for bricks, often extend below the water-table. As soon as pumping ceases, these pits become flooded. The floor of the workings is often very uneven, as the machinery which excavated the clay or gravel was

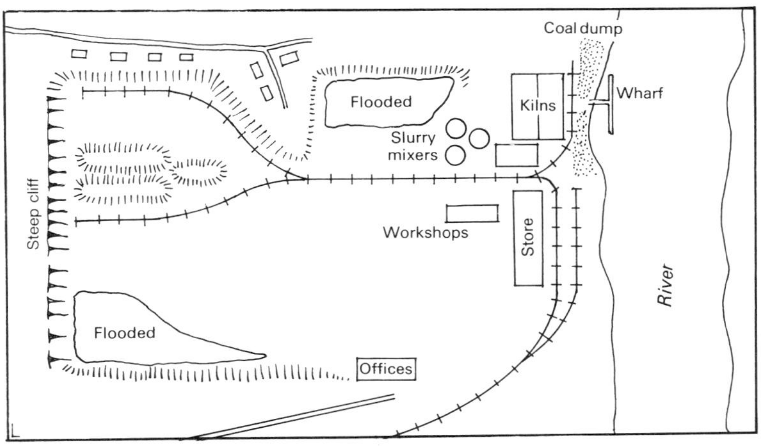

Fig. 3.9. Diagrammatic plan of a cement works.

Fig. 3.10. Active gravel pit, Trent Valley, Nottinghamshire.

used to return reject material within the confines of the quarry. Finer material resulting from washing and grading sand and gravel is passed into lagoons or settling ponds, forming areas of very unstable ground (Fig. 3.10). Clay pits for brick manufacture often contain reject and spoilt bricks as well as the overburden of glacial till. Situated immediately adjacent to the clay pits are the kilns and other buildings containing machinery for the manufacture of bricks and tiles. In recent years the kilns with their large chimneys have been located in the bottom of dry clay pits, but, wherever they are situated, their tall chimneys require substantial foundations. There may have been emissions of dust, fluoride, and sulphur dioxide during the operation of the works.

Many of the problems of restoration of quarries result from the use of these sites after the end of quarrying. Unauthorized fly-tipping of chemicals and other unpleasant materials is common, as well as the official dumping of household and other wastes.

EXPECTED PROBLEMS: acidity, buildings, domestic refuse, dust, flooded workings, lagoons, quarry faces, rock waste, rubble, lack of soil. (See also *Railways*, p. 32, and *Canals*, p. 33.)

Open-pit mining

This process has been separated from ordinary quarrying because of its size and complexity. Open-pit mining is used to extract low-metal-content porphyry copper ores in Utah and Arizona in the USA (Fig. 3.11). The ores are mined from terraces round large bowl-shaped pits which can be over a mile in diameter and over 1200 ft in depth. In addition to the immense size of the hole, much rock waste remains after concentrating the ore (Fig. 3.12).

EXPECTED PROBLEMS: buildings, dust, excavation, flooded working, lagoons, metallic contaminants, (see individual metallic elements listed in Chapter 4), quarry faces, rock waste.

Opencast mining

Opencast coal-mining has been described (p. 15). The Jurassic iron ores of Britain have been extracted using the same method. Initially, they were mined with wheelbarrow and shovels at the beginning of the century. Mechanization before the Second World War produced parallel ridges of waste known as 'hill and dale' which resulted in the Kennet Committee (1939) enquiry which could see no solution to the problem then. However, the development of earth-moving machinery during the Second World War made possible both deeper workings and their subsequent restoration (Fig. 3.4). Since 1951, restoration of opencast sites in Britain has been mandatory (Brent-Jones 1976).

Fig. 3.11. Bingham Copper Mine, Utah. The size of the operation and amounts of rock waste are very large.

EXPECTED PROBLEMS: acidity, colliery waste, combustible material, quarry faces, rock waste, soil erosion and lack of soil, subsidence.

Wet mining

Tin, gold, and certain precious stones are extracted from alluvial deposits in river valleys by dragline or floating dredge. The dragline or dredge takes the alluvial sediments from the valley floor, the metals or precious stones are separated, and the waste material is then returned to the valley floor. Normally, the amount of mineral extracted is less than 0.5 per cent so almost all the alluvium is returned, though less consolidated. These alluvial deposits are usually at or below the water-table, so site drainage is much disrupted. With an integrated dredge the pay minerals

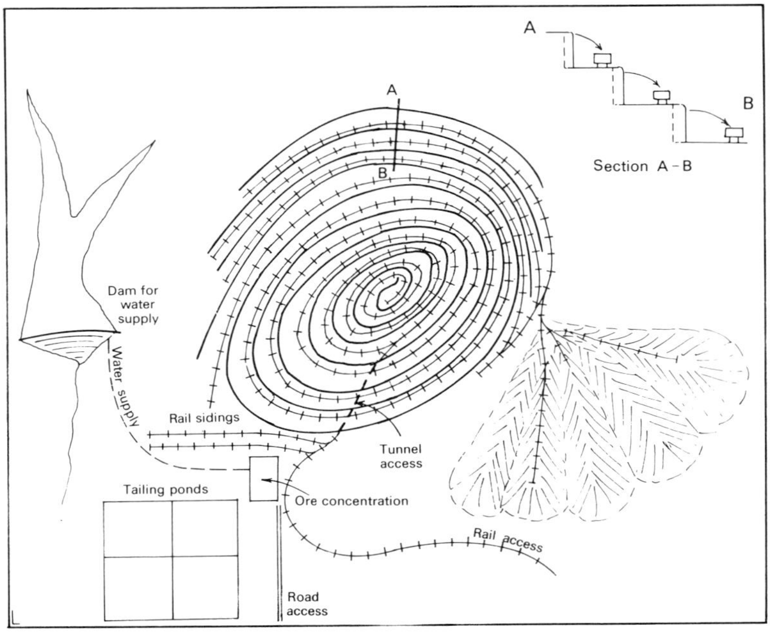

Fig. 3.12. Diagrammatic plan of an open-pit mine.

are separated on the dredger and the spoil is returned to the pit behind the dredger in a series of 'hills and dales'.

China clay is extracted by high-pressure hoses which wash the kaolin from the weathered granite. The clay is then separated from the quartz and mica; 10 tons of kaolin give 1 ton of clay and 9 tons of quartz sand. Kaolin mining produces a large hole, but since there is usually more kaolin present, the hole is not usually infilled with the waste, so the coarse waste is tipped alongside in the characteristic conical heaps seen in the St Austell district of Cornwall. The clay is separated from the mica in lagoons (Bradshaw *et al.* 1975).

EXPECTED PROBLEMS: acidity, dust, flooded workings, lagoons, quarry faces, rock waste, soil erosion.

Derelict land from metal smelting

The early metalliferous industries grew where it was most convenient to assemble ores, fluxes, and fuel together. Thus, copper and zinc smelters were built along the tidal reaches of the River Tawe in Swansea where the ore ships could be unloaded. Canals and railways were built to bring coal from mines further inland. The iron and steel industry began in wooded areas such as the Weald and in the valleys west of Sheffield where water power was available to work the metal. When the industry changed to coal for its fuel, the iron and steel works were built on the coalfields as at Ebbw Vale and Consett. The most recent change has been to move to a coastal location where iron ore can easily be imported as at Port Talbot and Ravenscraig. Until recent nationalization in the metalliferous industries, there were small metal-processing works in many parts of the country, any of which may become a derelict site with contamination persisting, even though the land may have been put to other uses subsequently. The surveyor of derelict land needs to investigate such sites carefully.

Smelting of non-ferrous metals. Many of the non-ferrous metals are smelted from their sulphide ores, e.g. copper, lead, zinc, cadmium, nickel, and arsenic. In the nineteenth century this mainly involved heating the ore in a furnace to drive off the sulphur as sulphur dioxide, and other impurities were taken off in the slag which floated on the surface of the molten metal. The sulphur dioxide emitted blighted the vegetation for about a mile around the furnaces. Once killed, the protective effect of the vegetation ceased and soil erosion began, so gullied bare ground with little vegetation is a common site around older non-ferrous metal smelter sites (Fig. 3.13). Outside the immediate zone of the smelter, vegetation may show signs of stress for up to 20 miles. Unfortunately, the smelter

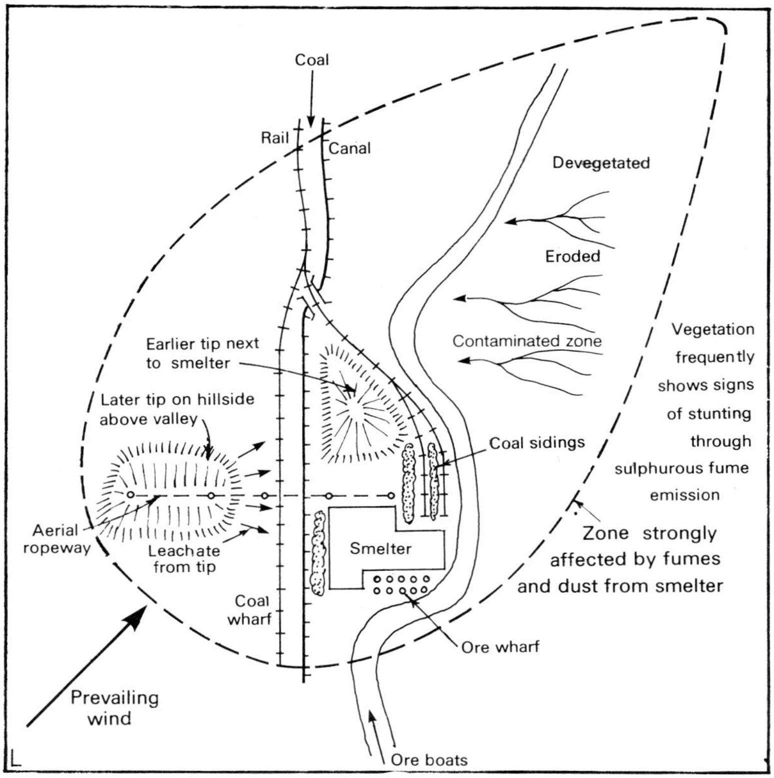

Fig. 3.13. Diagrammatic plan of nineteenth-century copper, lead, or zinc works (based upon Hafod/Morfa Works, Swansea).

owners turned to whichever metal was most profitable, so in most cases there is multiple contamination. Also, minor constituents like arsenic, cadmium, and nickel were released from the chimneys as oxide dust which settled in and around the site. Seepages from the slag heaps may contain metallic contaminants which affect the surrounding soils. The buildings which housed the smelters are usually fairly substantial and with their foundations offer problems for redevelopment. There may also be relics of railway sidings and coal wharves.

In aluminium production, bauxite ore is first digested with sodium hydroxide to bring the aluminium into solution as sodium aluminate from which aluminum hydroxide is subsequently precipitated. The iron-rich residue contains excess sodium hydroxide and is strongly alkaline. The iron-rich residue is mainly fine-grained, but it contains some material of sand size. The alkalinity of the sandy material can be leached fairly easily, but the red mud is impermeable and difficult to leach.

EXPECTED PROBLEMS: acidity, arsenic, cadmium, copper, lead, mercury, metal wastes, nickel, rubble, soil erosion, zinc.

Production of iron and steel. Metallic iron is smelted from its ores in a blast furnace with coke and limestone, which acts as a flux. High temperatures are achieved by passing hot air through the blast furnace charge. Liquid iron collects at the bottom of the furnace and a slag of impurities floats above it. Steel is made from the iron in a subseqent operation in which the impurities are oxidized and other elements added according to the variety of steel required. Steel is produced from several types of converter and from electric furnaces.

Much of the waste associated with late nineteenth- and early twentieth-century steel works came from Bessemer converters or open-hearth furnaces. The Bessemer converter is a pear-shaped furnace which can be tilted for charging, blasting, and pouring. The charge consists of pig-iron from the blast furnace together with the additions necessary for particular types of steel. The quality of steel produced by Bessemer converters is variable because the process is rapid and difficult to control accurately. The open-hearth furnace, developed in 1858 by Siemens in Swansea, provided a slower and more easily controlled process. It took about 12 h to produce

100–50 tons of steel compared with 15 min for about 50 tons in a Bessemer converter. Both have been superseded by electric furnaces, which, although smaller, can be more accurately controlled for high-grade steel production.

A process which combines the speed of the Bessemer process with the high quality of the open-hearth process is the Linz–Donowitz (LD) or oxygen method of steel production. Pig-iron is converted to steel by a blast of almost pure oxygen delivered into the molten metal by a water-cooled lance held just above the molten metal.

All furnaces are basic-lined now except in foundries. Acid (silica)-lined furnaces were used for about 150 years, for iron ores low in sulphur and phosphorus. Basic-lined furnaces have predominated for about 40 years. The basic lining is necessary to deal with iron ores containing more than 0.08 per cent phosphorus.

Following the building of the first continuous rolling mill in 1926 in the United States, steel works grew larger. A modern integrated iron and steel plant with its rolling mills can be almost a mile long with raw materials going in at one end and sheet steel coming out at the other (Fig. 3.14). The older steel and tinplate works which have been closed during the past 40 years occupied much smaller areas. Such works would have had a limited amount of tipping which added to the extent of derelict land when they closed. The smelting of iron ore needs much coke, so most iron works had a coke preparation plant (see *Carbonization plants*, p. 19), and some sites now present similar problems to gasworks sites.

Traces of nickel, chromium, beryllium, vanadium, tungsten, titanium, molybdenum, manganese, cobalt, and lead used in alloying metals may be present in small amounts on abandoned sites. Cleansing chemicals, acids, cyanides, degreasing substances, and other solvents may have been used in manufacturing processes performed upon the steel and could have contaminated the site.

A steel works may have been dismantled, in which case little of the above-ground structures can be seen, though their strong foundations and underground tanks and pipework may remain.

EXPECTED PROBLEMS: acidity, asbestos, buildings, combustible materials, metal wastes (see metal wastes and individual metallic elements listed in Chapter 4), iron and steel slags, rubble.

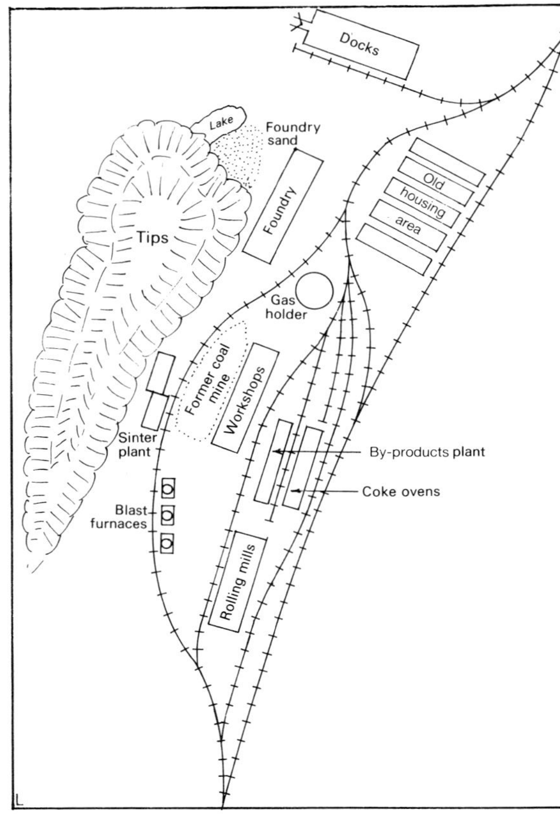

Fig. 3.14. Diagrammatic plan of an iron and steel works (based upon Workington, Cumbria).

Derelict land from the chemical industry

Land formerly occupied by the chemical industry may present a wide variety of contamination problems. Some contaminants weather rapidly to harmless materials; others are highly toxic and persist for long periods.

Until the end of the nineteenth century and the early part of the twentieth century, chemical industries were on a relatively small scale and had only a small environmental impact. However, some dumps of waste products (e.g. alkali and chromate wastes) remain as derelict areas, lacking vegetation, and are an unnecessary visual intrusion in the landscape. Where they weather these materials can sometimes produce interesting sites for specialized forms of plant life (Gemmell 1977; Greenwood and Gemmell 1978). Although a problem, wastes like these have gradually merged into the landscape. More toxic chemicals can be a hazard even when buried deeply.

Except for severe accidents such as that at Seveso

in Italy and at Flixborough in Lincolnshire which pollute the site of a chemical works and its neighbourhood, the main dangers from the chemical industry are in the disposal of non-saleable or useless by-products, and the persistence of contaminated equipment or building rubble when the works are dismantled. Recent legislation in many countries prohibits the indiscriminate disposal of toxic wastes, but previously many chemical plants disposed of their toxic wastes within the factory grounds by dumping solids in heaps and placing liquids in lagoons. Under present legislation toxic wastes may only be disposed of in Great Britain at sites registered by the Waste Disposal Agency, normally the local authority. The problem of waste chemicals will be returned to in *Derelict sites with landfill*, p. 33.

General chemicals. The DoE in the UK has indicated that when chemical works become derelict their sites are characterized by problems of contamination. At present, there is not a great deal of information available but national organizations in several countries are beginning to compile data upon these problems. Thus a sulphuric acid plant at Cutler, Ontario, ceased to operate in 1963 when its equipment and rail siding were removed. The buildings were demolished in 1969 and the land returned to the Crown, but problems occurred because pyrites remaining on the site were making the surface drainage strongly acidic. A chemical factory operated near Hamburg between 1929 and 1979; the complete history of its activities is not known, but between 1944 and 1979 it produced lac. A concentration of 48 750 mg/kg of cyanide was found in the soil at 0.5 m depth, and 28.5 mg/l was present in the ground water. Another factory in the Hamburg area was found to have remnants of chemicals used to manufacture poison gas during wartime. Also in Germany, a chemical factory site at Gendorf, Bavaria was found to contain 400 tonnes of HCH (hexachlorcyclohexane) a by-product from the manufacture of lindane between 1949 and 1954. The site had been graded in the early 1960s and covered with 0.5 m of soil. No contamination was found outside the site which has now been encapsulated.

Accidental leakage from underground tanks or pipes may lead to site contamination which remains after the factory has been dismantled. An example of leakage occurred at Regina, Canada where 21 000 l of a liquid containing 70 per cent PCBs and 30 per cent TCBs escaped. Although most of it was collected and removed, great concern resulted about contamination of the ground water (Smith 1985).

Arsenic was used in the USA during the 1930s and 1940s to control a plague of grasshoppers. When the control programme ceased, left-over arsenic was buried on site at two locations in Minnesota where the poisoned bait was mixed. Eventually, the buried arsenic led to soil and ground water contamination and people were poisoned after drinking from local water supplies. In Britain a leakage of ammonia chrome sulphate occurred from a photographic chemical factory, a by-product of hydroquinone manufacture. Subsequently, an explosion at the same site released dimethyl ether and sulphuric acid, and dimethyl sulphate was produced which was found remaining in the soil 13 years later.

Former man-made fibre plants are being demolished at Flint in North Wales where the ground within the works has been found to be contaminated. Hydrogen sulphide was present in the soil and a tank formerly containing 25 000 l of carbon disulphide was found and had to be removed. Sulphuric acid used in the manufacture of rayon had caused corrosion of brickwork, and sodium hydroxide was suspected as another contaminant elsewhere.

Derelict sites where paints were formerly manufactured are also found to have problems of contamination. The paint industry generates approximately 135 million l of solvent waste each year, and uses toxic metal salts as pigments together with resins, blenders and preservatives, many of which are also toxic materials (DoE 1977*e*). Inevitably, sites are found to be contaminated as correct disposal of wastes has not always taken place. The production of titanium dioxide, used instead of lead, results in acid iron wastes.

Sites where explosives have been manufactured, stored, and disposed of may contain several contaminants. Many arsenal sites have been released by the Ministry of Defence for alternative uses. The Woolwich Arsenal on the south bank of the Thames is the best known site, but there are others such as Pembrey near Llanelli. A new suburb is being built at Thamesmead, and the Pembrey site has become a Country Park. Potentially explosive chemicals may occur in landfill sites, but since 1974 it has been mandatory to notify 'special wastes' and dispose of them at specified sites only (Control of Pollution Act 1974). Explosives are made from nitrogen and phosphorus compounds, so a nitric acid plant is a common accompaniment of a munitions factory. As strategically important works, they often had their own gasworks and electricity supply on site, so production would not cease if public supplies were disrupted.

Until the development of rocket-fuelled missiles and atomic warheads, the most commonly used explosive materials were gunpowder which is a mixture of potassium nitrate, charcoal, and sulphur, gun-cotton or trinitrocellulose, which is produced by the action of a mixture of sulphuric acid and nitric acid on cotton, and nitroglycerin, which is produced by the action of the same acids on glycerin; nitroglycerin is then absorbed in diatomaceous earth (kieselguhr) to form a greasy, plastic solid, dynamite. If ignited, dynamite will burn quietly but if detonated in a small space it has great shattering power; mercury fulminate is used as a detonator. Cordite is a paste of nitroglycerine gun-cotton and vaseline in acetone. After being forced through a die to give thin strands, the acetone is allowed to evaporate, leaving the brown cordite. Picric acid or trinitrophenol was used for high explosive shells, but trinitrotoluene was safer to handle. Amatol is composed of trinitrotoluene and ammonium nitrate and potassium chlorate is used to make cheddite.

Sites like the Woolwich Arsenal have been used for a long time. Often they began as foundries casting cannon in iron or brass. The manufacture of explosives accompanied this metallurgical work and up to 56 000 people were employed at Woolwich during the First World War. From 1860 the Woolwich Arsenal had its own town gas supply and an electricty generating station was added at a later date. Although vulnerable to air attack, armaments production continued at Woolwich throughout the Second World War. After the war, large quantities of explosive materials were disposed of on the site by burning and the residues were added to the waste materials already dumped. Disposal of military flares leaves traces of copper, nickel, cobalt, strontium, calcium, and barium. Smoke bombs leave behind zinc oxides and hexachlorethane and residues of TNT and fulminate of mercury and graphite frequently occur (Lowe 1980).

In the future, sites used for weapons manufacture will also need to be checked for radioactivity; a recent leakage at a Royal Ordnance Factory has released beryllium dust stated to be used for packing around nuclear warheads in Trident missiles.

Many of the contaminants found on the sites of former chemical works are hazardous and must be treated with great caution. Information on these chemical wastes from a wide range of industrial activities is given in Waste Management Paper 23 (DoE 1981) where several hundred compounds are listed with their critical characteristics such as toxicity, flash-point, corrosivity, and carcinogenicity. Table 3.1 indicates a logical approach to the assessment of the status of these hazardous waste materials.

EXPECTED PROBLEMS: acidity, alkalinity, arsenic, asbestos, cadmium, chromium, copper, cyanide, dioxins, dust, halogenated hydrocarbons, lagoons, lead, mercury, nickel, paints, pesticides, phenols, polychlorinated biphenyls, rubble, solvents, sulphates, zinc. (See also *Carbonization plants,* p. 19; *Thermal power stations,* p. 17; *Railways,* p. 32; *Scrapyards,* p. 32; and *Derelict sites with landfill,* p. 33.)

Metal goods, engineering, and vehicle industries

The pattern of dereliction and contamination from engineering works will be as diverse as the activities they carried out. Engineering works may be of any size from a small back-street enterprise to a fully integrated manufacturing and assembly works. Heavy engineering works deal with constructional materials and the wide range of machine tool manufacturers and other light engineering works deal with an assortment of metals and processes. Potential contaminants on former engineering works sites may be liquids, sludges, and solids.

Wastes from metal-processing techniques such as cleaning, degreasing, electroplating, anodizing, galvanizing, rust-proofing, machining, and polishing may have been disposed of on site. They may contain copper, chromium, lead, zinc, nickel, cadmium, mercury, aluminium, or iron as well as noxious solvents. Many of these substances come into the category of 'special wastes' described in the previous section. Since the Control of Pollution Act 1974, these wastes have been considered to be 'special wastes' and must be disposed of only at specially licensed sites.

Mineral oils are widely used for lubricants and the cooling of machining and drilling operations. In many cases it is possible to recover the oil from the 'swarf' and to use it for other less demanding purposes, but there comes a point when the remaining sludge and its contaminants must be disposed of. Oily wastes may pose a risk of potential combustion below ground, and they can also cause skin irritation or even cancer if a sufficiently long exposure is maintained. Some oils contain anti-oxidants, anti-knock agents, and bacteriocides and should be handled with care.

Solvents, such as paraffin, white spirit, and paint

Table 3.1.

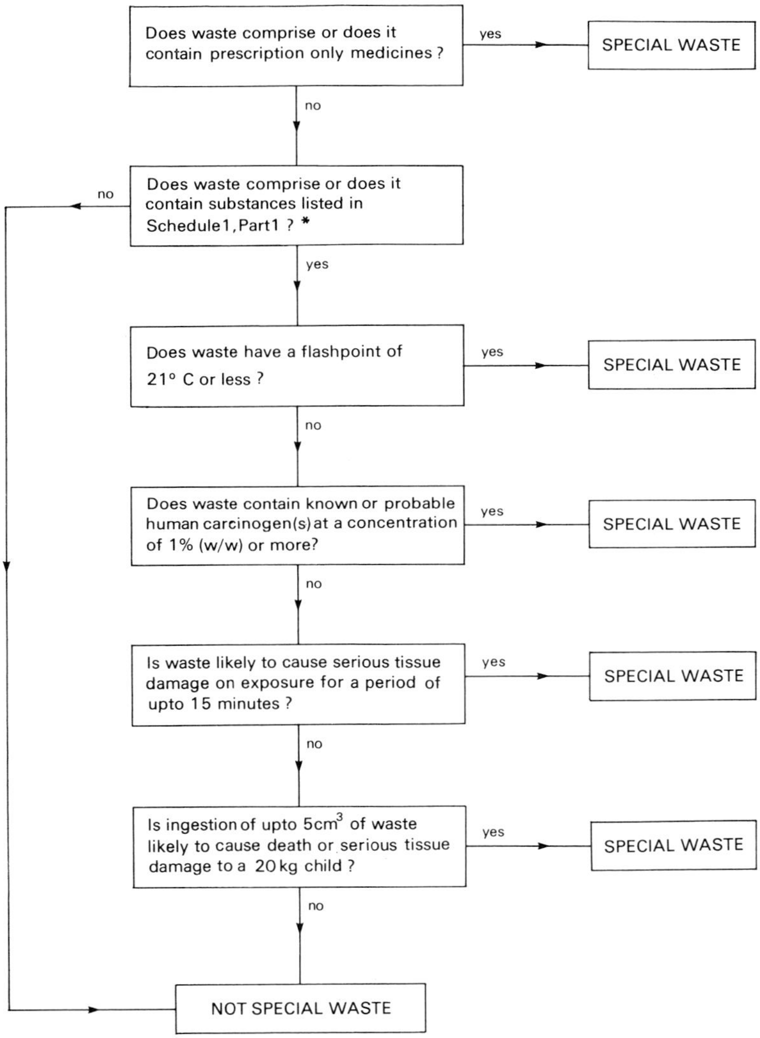

[This table is Crown Copyright and is reproduced here by permission of the Department of the Environment, from Special Wastes, *Waste Management Paper No. 23.* DoE 1981.]

*Substances which (a) are dangerous to life, (b) have a flashpoint of 21°C or less, or (c) are medicinal products defined in Section 130 of the Medicines Act 1978.

thinners, and degreasing agents like phenols and cresols used in metal-cleaning operations before painting or electroplating are also likely contaminants of an engineering site. Engineering works which have refitted or dismantled ships or railway carriages insulated with asbestos should be inspected with great care as asbestos dumped on site requires special treatment defined under the regulations concerning its handling.

EXPECTED PROBLEMS: asbestos, combustible materials, lagoons, metal wastes (see metal wastes and individual metallic elements listed in Chapter 4), metal waste liquids, oil, paints, phenols, rubble, and solvents.

Derelict land from other manufacturing industries

Although the amount of derelict land caused by the

industries included in this category is small compared with the potential dereliction of the metalliferous industries, the sites are often in, or close to, residential areas or areas where redevelopment is taking place.

Printing works

Until recently, the small jobbing printer used metal type which was reset for each new job required by customers. Larger and more modern printing operations, including letterpress and newspaper printing, generate wastes which may pollute the sites of their former works. In addition to solvents used for cleaning, such as toluene, benzine, and phenol, site contaminants may include chromic acid and other strong acids used in the etching of printing plates. These spent etchants may contain metals such as copper, chromium, lead, and zinc.

Printing ink manufacturers have similar wastes and produce similar site pollutants to those of the paint industry but in smaller quantities. A survey of a former printing ink manufactury should check for cadmium, chromium, copper, lead, nickel, zinc on the site (DoE 1977*e*).

EXPECTED PROBLEMS: acidity, chromium, copper, cyanide, lead, nickel, phenols, rubble, solvents, zinc.

Tanning, leather, dressing, and fellmongering

Although the materials of this industry are biodegradable, some of the chemicals used in processing are toxic and may remain as potential pollutants on a derelict site or in abandoned factory premises. A former tannery site will contain buildings with tanks and pickling baths below ground level which may present problems and additional expense prior to redevelopment.

During pre-tanning treatments skins are soaked in aqueous solutions to remove impurities such as dirt, extraneous tissue, and hair. Sodium hypochlorite is used to destroy anthrax spores; lime and sodium sulphide are used to remove tissue and hair. Treatment with acid ammonium salts removes excess lime. Tanning was formerly accomplished with natural tannins, but nowadays skins are preserved with chromium (III) sulphate. Finishing processes employ solvents, dyes, and pigments containing toxic elements. Wastes from a tannery have a high biological oxygen demand estimated to be ten times that of domestic sewage, and, with the contained chromium and other salts, are a potential hazard to surface watercourses (DoE 1978*b*).

The by-products of a tannery include trimmings with grease, tallow, and gelatine which are extracted from the tanned and untanned hides and off-cuts. Similar wastes arise in slaughterhouses and glue factories where bones and fleshings are reduced to gelatine and glues.

EXPECTED PROBLEMS: buildings, chromium, rubble.

Wood-processing industries

Although wood is biodegradable and non-toxic it is perishable so that various toxic materials are used as preservatives and to prevent insect and fungal attack.

Wood chippings, shavings, and sawdust will have become incorporated into the soil and enhanced its organic content. It is conceivable that the amounts could be sufficient for the soil to be combustible.

However, the insecticides and fungicides used for preservation are likely to be the more dangerous materials left on a site. Wood preservatives fall into three categories: tar oils derived from the carbonization of coal (see p. 19) consist mostly of creosote which contains a high proportion of aromatic hydrocarbons and a small proportion of phenolic derivatives; organic solvent-based compounds consist of one or more organic compounds such as pentachlorophenol (PCP), gamma hexachlorocyclohexane (HCH) (formerly known as gamma benzine hexachloride (BHC)), metal naphthenates, bis (tributylin) oxide (TBTO), or dieldrin dissolved in an organic solvent; copper 8-quinolinolate, pentachlorophenyl laurate, and 2-phenylphenol are less commonly used. The most common water-based preservative is a copper–chrome–arsenic mixture (CCA) containing copper sulphate, sodium dichromate, and arsenic pentoxide (DoE 1978*a*).

Telegraph poles are usually treated with creosote and have a useful life of about 40 years, but some are treated with copper–chrome–arsenic mixture. The wooden sleepers of British rail were also treated with creosote but now these have been extensively replaced with concrete ones. The inside of the cooling towers of the CEGB contain between 250 and 350 tonnes of wood splash-packing which has been copper–chrome–arsenic-treated to combat deterioration in warm wet conditions.

Toxic wastes may arise on wood-working sites where these substances are manufactured or formulated, or where wood is treated, where preservative containers have been carelessly disposed of, or where there has been an accumulation of waste wood, shavings, or sawdust from timber being worked after impregnation.

EXPECTED PROBLEMS: arsenic, buildings, chromium,

combustible materials, chemical wastes. (See *Coal carbonization*, p. 19.)

Rubber and synthetic rubber factories

Natural rubber is obtained from the sap of the rubber tree, *Hevea braziliensis*. The sap, or latex, collected from cuts in the bark, is coagulated, squeezed, and cured in a smoky atmosphere. Latex may also be stabilized with ammonia and transported to factories as a liquid. As a natural product, rubber is subject to biodegradation and unlikely to form a site contaminant.

On arrival at the factory, the raw rubber is masticated to make it soft and putty-like before other ingredients are added. Small amounts of sulphur (8–10 per cent) give rubber a soft flexibile consistency, but with greater amounts (30–40 per cent) it forms a hard brittle substance after heating (vulcanization). Other ingredients for motor tyres are carbon black, pine tar, anti-oxidants, zinc oxides, stearic acid, mercaptobenzothiazoles, and diphenyl-guanidine. Rubber can be dissolved by benzene, carbon tetrachloride, and naphtha, all of which are unpleasant chemicals and may be present on a site formerly occupied by a rubber factory.

Synthetic rubbers have been produced from organic chemical precursors such as styrene and butadiene. They have similar properties to natural rubber and can also be vulcanized. In their finished form, both natural and synthetic rubbers are relatively benign but persistent. If burnt, some of the toxic intermediary compounds may be released, together with carbon.

EXPECTED PROBLEMS: solvents, zinc.

Demolition and construction

The rapidly extending area of derelict land associated with the inner parts of many cities presents a new aspect of derelict land which has become increasingly apparent in the last 20 years. These areas were formerly occupied by housing for working people which was often poorly built between 100 and 150 years ago. It was within easy walking distance of small industrial concerns but now both housing and industrial buildings have reached the end of their useful life.

Redevelopment of these inner city areas requires large capital investment and it is often many years before landholdings can be consolidated by the local authority so that redevelopment can begin. In other cases companies or persons have held land as an investment against inflation after the property upon it has been demolished. Although a perfectly legal procedure, it does maintain land in a derelict state.

In a study of the urban environment of the West Midlands County, Haines (1980) revealed that there is a build-up of heavy metals associated with industrial conurbations. A central area of West Bromwich had higher mean metal levels when compared with a mixed industrial/housing area (Langley) and an interwar suburban residential area (Halesowen) (Table 3.2). Comparing these figures with background values from uncontaminated rural soils, lead levels were definitely increased but cadmium levels were not abnormally high. The figures broadly support similar findings contained in surveys carried out in South Wales for the Welsh Office (Goodman *et al.* 1975).

Industrial and manufacturing premises will have their specific problems arising from the former uses of the site or buildings. These sites must be treated as special cases and care taken to isolate any areas with abnormal problems. Within former residential areas, the levels of toxic elements are not likely to be excessive but problems include rubble contaminated with gypsum and lead paints. Cellars and underground services also present potential difficulties during site redevelopment.

EXPECTED PROBLEMS: asbestos, buildings, combustible materials, land ownership, lead, rubble, sulphates.

Table 3.2. Levels of lead, cadmium, and copper in urban soils of the West Midlands

	Available (µg/g dry weight)			Total (µg/g dry weight)		
	Pb	Cd	Cu	Pb	Cd	Cu
West Bromwich	105.6	1.2	211.4	278.3	3.2	403.3
Langley	54.6	0.82	35.3	321.9	3.35	195.8
Halesowen	28.6	0.33	6.4	75.6	2.6	40.9

Distribution, hotels and catering, repairs

A wide range of service industries is included in this category. Dereliction resulting from retail distribution, hotels, and catering facilities presents similar problems to general urban decay. The major contaminating land use in this category is scrap-yards, dealing with metals and other recyclable wastes.

Scrap-yards

Any land which has been used as a scrap-yard almost certainly contains problems for redevelopment. The ground surface in any former scrap-yard will inevitably include a layer of particulate metal fragments and will be soaked in oil. Other solid, liquid, and sludge wastes will almost certainly have been disposed of on site. As a result of the interaction between wastes there may also be toxic gas emissions from the ground.

Although iron is usually the most common metal in scrap-yards, it is associated with many other elements, some of which are harmful at any concentration. The concentration of arsenic, antimony, cadmium, chromium, copper, iron, lead, manganese, mercury, nickel, tin, and zinc may be high enough for concern (ICRCL 1980).

Although 'normal' levels of trace elements in uncontaminated agricultural land and derelict sites may be compared, it is still not possible to generalize about what levels are acceptable on any one site. Tentative suggestions have been given (ICRCL 1983) and the problem continues to be under review.

Other toxic substances which may be present on scrap-yard sites are polychlorinated biphenyls (PCBs) from broken or dismantled electrical equipment. These substances may be absorbed through the skin and once in the body do not break down. If PCBs are passed along the food chain until they reach human beings they may cause cancer of the liver, deformities in babies, disfiguring skin complaints, and vomiting. Obviously, if any of these noxious chemicals have been stored in tanks or drained into the ground on a scrap-yard, there could be a problem for redevelopment.

Other materials, which may be present on the sites of former scrap-yards which can be potentially hazardous if the site is disturbed in redevelopment, include dilute or concentrated acids derived from lead-acid batteries, derusting and anti-fouling paints; and heavy metal dusts arising from burning of lead–acid batteries to obtain the lead; some types of oil containing phenolic substances may be encountered and other organic solvents may occur. Asbestos used for insulation purposes in vehicles is a hazard which may occur at any site where vehicle dismantling (road, rail, or sea) has taken place.

A further problem with reuse of scrap-yard sites stems from the practice of disposing of waste oil into lagoons or haphazardly into the ground on site. As a result the soil may reach the point where it will burn if ignited. It is suggested that any contaminated soil with a calorific value of 10.0 MJ/kg or more may represent a hazard from combustion (ICRCL 1984).

If buildings are proposed upon a former scrap-yard site the posibility of sulphate or acid attack on foundations and service pipelines should be investigated. The junction 'O' rings of pipes, and certain plastics, may be corroded rapidly and lead to leakages and possibly contamination of water supplies. The emission of gases is also a potential hazard on scrap-yard sites. Methane, hydrogen sulphide, and carbon dioxide may all accumulate in basements or enclosed spaces of adjacent buildings or structures built upon sites. Piling for buildings may provide access for solutions or gases to migrate through the soil to reach aquifiers; consequently, the geology and hydrogeology of any site scheduled for redevelopment should be carefully assessed. Many of the problems of scrap-yards are similar to those of domestic refuse discussed in *Derelict sites with landfill*, p. 33, which should be read in conjunction with this section. In a limited way, redevelopment of an old motor garage repair shop may give rise to similar problems of ground contamination.

EXPECTED PROBLEMS: acidity, asbestos, combustible materials, dust, gases, lagoons, metal wastes, oil, paints, phenols, polychlorinated biphenyls, rubber, solvents, sulphates.

Transport and communications

As the pattern of industry changes, the transport services which serve industry are forced to change and, where contraction takes place, land becomes derelict. The greatest changes have been in the canals and railways, which have contracted greatly during the past 50 to 100 years. In both cases a linear strip of land is abandoned with more extensive areas where former sidings, stations, or wharves were located.

Railways

In many country districts it has proved possible to incorporate linear sections of former railway land into the adjacent fields. Where a former canal or railway ran parallel to a modern busy road, the track has often been utilized as a second carriageway for the road. Some railways have become cycle tracks or walkways; some former cuttings have been infilled with domestic refuse.

The physical problems associated with railway land which has become derelict are associated with its narrow width, the steep sides of cuttings and embankments, the presence of ballast from the track bed, old sleepers, and rails (although the metal and wood are normally removed by scrap merchants). There is a possibility of combustible material at engine depots where boilers were raked-out. The ash included incompletely burnt coal which, with the reject coal dust, has left a combustion hazard, notably at Swindon (see p. 119). A large railway works, such as Swindon, could also have problems associated with metal-working and carriage works could have handled asbestos, some of which may remain buried on site.

EXPECTED PROBLEMS: asbestos, buildings, combustible materials, metal wastes, oil, paints, rock waste (ballast), rubble.

Canals

The problems of canals are similar to those of the railways in that a long narrow strip of land has been abandoned. There may be tunnels but canals usually followed the contour of the land. Flights of locks can leave derelict brickwork or concrete with underground tunnels through which water was transferred in and out of the locks. The drainage around an abandoned canal is usually poor as maintenance has long since ceased. Silted canals and basins can cause trouble under building works as settlement occurs on loading the unconsolidated sediments, and methane gas is generated from anaerobic breakdown of organic matter.

Sludge accumulates on the puddled floor of the canal and its basins. This organic-rich deposit may have been contaminated by material dropped from boats or allowed to drain into the canal waters. Edwards and Pumphrey (1980) report a pH of 9.2–9.5, salinity, and elevated mercury levels in canal dredgings where effluent from a chlor-alkali works had been discharged.

EXPECTED PROBLEMS: acidity, alkalinity, combustible material, gases, metal wastes, rubble.

Docks

Many docks were excavated from poorly consolidated sediments and the excavated material used to raise the land level alongside the dock. In order to facilitate loading and unloading, the sides of docks are usually supported by substantial walls of masonry. Some of these may be up to 100 years old. Whilst submerged these are capable of supporting the dock sides. However, if the dock is drained, leaving saturated material behind the unsupported walls, they may collapse. As the depth of former docks may have been as much as 20 or 30 feet the dock walls present a precipitous change in level which can be dangerous and also present difficulties in site development.

A layer of sludge composed of organic and other debris has gradually accumulated on the bed of docks. Inevitably some spillage of toxic cargoes will have occurred and this may be sufficient to prohibit the use of dock sludge as a soil amendment. Mud from the London Docks contained enough lead to make it unsuitable for topsoil (Travis, Morgan, and Partners 1973).

The low-lying land on which docks were constructed may have been poorly drained with a high content of organic matter. Buildings placed on such sites should have adequate ventilation as methane may be produced, which can reach explosive proportions or can be asphyxiating in enclosed spaces. There may also be the danger of underground fire if the water-table is lowered and the site completely dries out (Carpenter 1986).

EXPECTED PROBLEMS: combustible materials, gases, lead, rubble. (See also *Railways*, p. 32.)

Banking, insurance, and other services

Banking and insurance services do not normally produce derelict land but 'other services', which include collection and disposal of domestic and industrial wastes and sewage are activities which can lead to land with contamination problems which remains in a derelict state.

Derelict sites with landfill

Domestic and industrial wastes are frequently used as landfill in formerly derelict pits and quarries; once full, the site may be reclaimed and brought back to productive use. Under the Control of Pollution Act 1974, every local authority in Great Britain is responsible for providing a system for the disposal of domestic and industrial wastes. Unfortunately, control of tipping was less rigorous before 1974 and many older landfill sites may contain wastes which could be harmful. However, in spite of some difficulties, there exist many examples of land successfully restored from landfill to productive use.

Currently, landfill sites are controlled by legislation under the Control of Pollution Act 1974, which in turn drew its recommendations from a Report of the Working Party on Refuse Disposal (DoE 1971). The accepted procedure is for the waste to be dumped in compartments, preferably on a puddled clay floor, in layers which are level and no more than 2.5 m deep. Each layer should be compacted and then covered as soon as possible with subsoil material. Finally, the specification recommends 1 m of soil material to cover the site (DoE 1976*a*).

The UK produces over 16.0 million tonnes of domestic waste a year, mainly composed of organic materials such as paper, wrapping materials, kitchen and garden waste, together with ashes, pottery, glass, and metal containers. Most of this material is non-toxic and represents approximately 350 kg per person per year. There are, in addition, some 23 million tonnes of industrial waste, 3 million tonnes of building waste, and 12 million tonnes of power station waste a year, most of which is disposed of by private contractors.

Since the Deposit of Poisonous Waste Act 1972 and the Control of Pollution ('special waste') regulations 1980, many toxic waste materials must be disposed of only at specially licensed sites. The EEC Council Directive on Toxic and Dangerous Wastes Brussels 1976, suggests that there be a cessation of co-disposal of domestic and chemical wastes. Experiments in America, Australia, and Canada endorse this view. However, in Britain toxic wastes may be 'co-disposed' with non-toxic wastes as investigations have shown that many of the metal contaminants are effectively held within the landfill and, where they escape into permeable rock beneath, they are strongly adsorbed. Sludges incorporated into an experimental landfill in aerobic and anaerobic conditions showed little change over a period of three years (Sumner Report 1978; Cope *et al.* 1983). Where these heavy-metal sludges are likely to come into contact with ground water or sensitive aquifiers, they must be treated with a chemical fixative to stop any migration.

If a landfill site in a permable rock is at, or below, the ground water table, there will be rapid transfer of toxic material into the rivers and pollution will result. Where there is an unsaturated zone below the landfill, there is close contact between the rock and the toxic leachate plume, and there is greater leachate attenuation than in saturated conditions (Aspinwall 1976; John and Davies 1976; Mather and Bromley 1976; Sumner Report 1978) (Fig. 3.15).

The Department of the Environment has reviewed the origins and methods of dealing with halogenated organic compounds, cyanides, metals, phenols, and phenolic substances, pigment wastes, tarry wastes, mercury, arsenic, polychlorinated biphenyls, acid wastes, mineral oils, and solvents, etc. in a series of Waste Management Papers (DoE 1976*a–i*, 1977*c–f*, 1978*a–d*, 1980, 1981).

Many of these toxic chemical wastes are sealed into drums before being consigned to landfill, but inevitably the containers rust and the chemicals are slowly released to become a hazard. Schuuring (1981) reported houses at Lekerkerke near Rotterdam built upon a waste dump containing materials from the dyestuffs industry including drums of toluene, xylene, and compounds of cadmium, lead, and zinc. Three hundred houses had to be evacuated and 1.5 million tonnes of polluted soil removed. In another case, infertility of cattle and the death of birds led to the discovery of a chemical dump in the Volgermeerpolder near Amsterdam which contained some 100 000 drums of the chemicals 2,4,5-T and 2,3,7,8-TCDD (Schuuring 1981). The factory which produced these by-products had been dismantled and dumped in the Atlantic, but the wastes remained. The Dutch government is said to have estimated that there are about 500 dangerous chemical dumps in the Netherlands alone which may cost £400 m. to clear and make safe.

The persistence of certain dangerous chemicals within landfill sites makes them potentially hazardous places should they be disturbed in the future. Although only certain sites are prescribed for toxic wastes, virtually all tips are suspect because of the activities of unscrupulous firms, or irresponsible employees, who have fly-tipped dangerous wastes in unauthorized places.

The erection of buildings on former landfill sites can present problems, some of which can be forecast from a survey. The risks include subsidence, underground fires, and the accumulation of gases in unventilated parts of buildings. All buildings should be piled but piling may provide access for polluted waters or for the entry of gases. Foundations of buildings and underground services are liable to be attacked and corroded on landfill sites. Acid and sulphate attack on concrete and the corrosion of metal pipes and plastics may eventually allow polluted waters into underground services. The

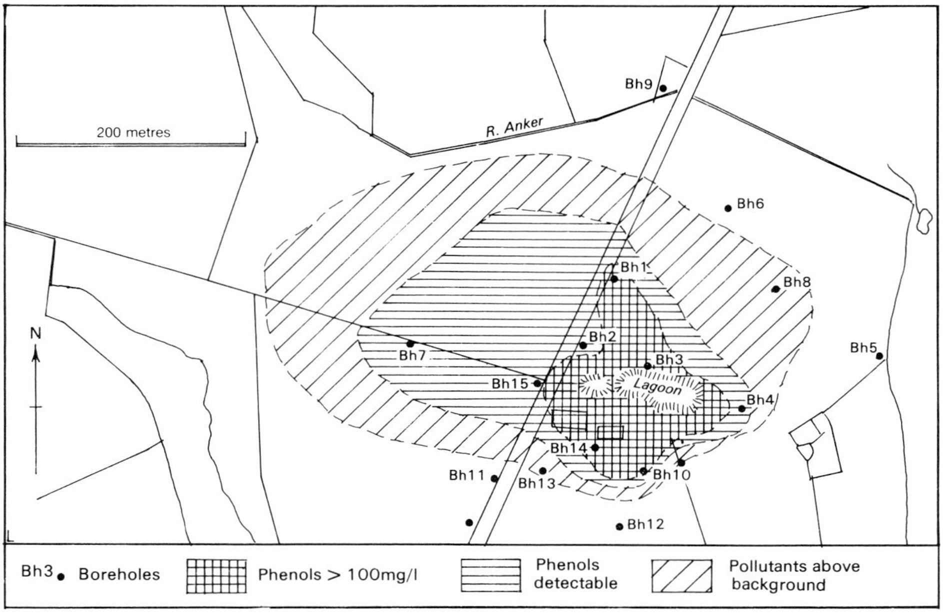

Fig. 3.15. Map of leachate plume from a landfill site in sandy material. (Crown Copyright. Reproduced by permission of the Department of the Environment, from Sumner Report 1978.)

British Standards Code of Practice for Foundations (BSI 1983) states that 'all made ground should be treated as suspect because of the likelihood of extreme variability'. Although documentation of sites is required now and the position of hazardous wastes marked on site plans, this sort of information does not always exist for sites operated before the present legislation came into force.

Where derelict sites containing landfill material are to be assessed for full reclamation the surveyor should be aware of the possibility of the numerous hazards and plan the survey accordingly (ICRCL 1978*a*).

EXPECTED PROBLEMS: acidity, asbestos, chemical wastes, combustible materials, domestic refuse, gases, lagoons, metal wastes, odours, pathogens, rodents and insects, rubble, subsidence, sulphates.

Derelict sites of former sewage works

The policy of the water authorities since 1974 has been to concentrate sewage treatment at centralized works of large capacity. As a result, many smaller sewage works have become redundant and may figure as derelict land. In rural areas, the reorganization of sewage collection from cesspits has reduced the need for 'sewage farms' which were formerly used for disposal of night-soil.

Treatment of sewage at a sewage works is a multistage process. The larger, floating debris is first removed and stockpiled. After further screening, the liquid and suspended matter passes on to sedimentation tanks. The suspended matter is allowed to fall to the bottom of the tank from which it is removed as raw sludge. The 'settled slurries' and the sedimented raw sludges are subjected to biological purification. Biological filtration takes place through filter beds constructed of coke, clinker, or foamed blast furnace slag. The liquid waste is supplied to the top of the filter bed and, as it trickles through, it is acted upon by the microbiological population which breaks down organic constituents and purifies the liquid in the process. Alternatively, the activated sludge process depends on micro-organisms in aerated tanks to break down organic matter; however, sufficient of the activated sludge must be recycled to keep the process going. In some works, the effluent from these processes is lagooned, filtered through sand beds or irrigated on to grassland to complete its purification. The sludge is air-dried, or in modern works undergoes anaerobic digestion to give a less obnoxious product, which can be sold as an organic manure (ICRCL 1979*c*).

There will be relatively few physical structures remaining on a former sewage works but there may

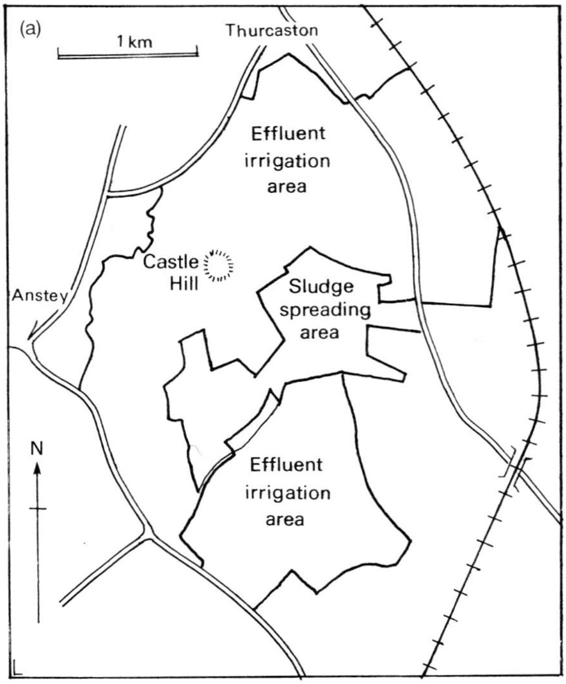

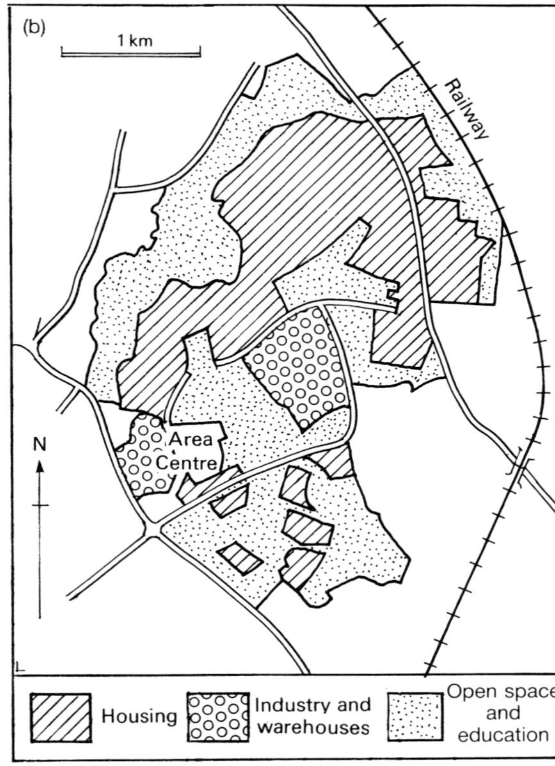

Fig. 3.16. Beaumont Leys. (a) As a sewage farm, (b) development plans.

be remains of pump houses, channels, and pipework associated with the transfer of effluent from one part of the site to another. There may be remains of old filter beds, lagoons, settling tanks, effluent clearing tanks, sludge-drying beds, and storage areas, as well as extensive grassed areas used for effluent dispersal. Some of the debris removed from the sewage in the initial screening may still remain on site and some of the lagoons may contain semi-solid or even liquid materials below a solidified crust. Sewage sludge may contain a wide variety of elements, toxic and beneficial, and may include, zinc, copper, iron, molybdenum, manganese, lead, cadmium, mercury, nickel, and arsenic.

A former sewage farm at Beaumont Leys Estate, Leicester which operated between 1890 and 1964 has been intensively studied. The land was required for urban development, so there was a site investigation. The site was sampled to 0, 2, 4, and 6 ft on a 200-ft grid and a wide range of total and available metals were determined. The farm could be subdivided simply into those areas where sludge had been spread and ploughed in (80 ha), and areas where the liquid effluent had been disposed of by irrigation (Fig. 3.16(a)). The irrigated area had been productively farmed with cereals, roots, and green vegetables and grazed by sheep, cattle, and flocks of turkeys. After 1964 it had been cropped with cereals and potatoes. Experimental evidence suggested that except for zinc no abnormal amounts of metals were being taken up by plants, though the sludged area exceeded the DoE trigger concentration for zinc equivalent. As a result this was designated as an area of public open space and the other areas were used for building (Heeps and Pike 1980; Wood and Ross 1980) (Fig. 3.16(b)).

EXPECTED PROBLEMS: remains of buildings, pump houses, channels, tanks, and pipework; combustible materials; gases; lagoons; metal wastes; odours; pathogens.

4 Components of derelict and contaminated land

Using a chemical analogy, the patterns of dereliction described in the previous chapter represent the 'molecules' of derelict and contaminated land, each presenting several problems or hazardous contaminants. Just as a small number of elements can form a large variety of molecules, individual components of the derelict land problem occur in many combinations. This chapter discusses a selection of the 'atoms' of the dereliction problem.

The Interdepartmental Committee on the Redevelopment of Contaminated Land (ICRCL) of the Department of the Environment has published guidance on the acceptable levels of certain contaminants which present hazards (ICRCL 1983). Its suggested values are tentative and must be interpreted according to site conditions (e.g. soil texture and drainage) and the proposed use after reclamation. The Committee introduced the concept of a 'trigger concentration' below which it is reasonable to consider a site as uncontaminated. When values exceed the trigger concentration, the site surveyor must decide what further investigation or remedial action is necessary. Certain elements have been claimed to have an additive effect on vegetation, e.g. copper, zinc, and nickel whose combined effect is described by the zinc equivalent (2Cu+Zn+8Ni), but this assumption is not universally accepted (DoE and National Water Council 1979). The Ministry of Housing, Planning, and the Environment (Netherlands) (1983) and Moen *et al.* (1986) have published figures for acceptable levels of contaminants, levels which indicate that further investigation is necessary, and levels at which reclamation measures are required (Table 4.1). The figures are broadly in line with the ICRCL figures (Table 4.2).

The relative hazards of many common flammable, explosive, and toxic substances used in chemical laboratories can be found in handbooks such as *Hazards in the chemistry laboratory* (Muir 1977), *Handbook of laboratory safety* (Steere 1967), and *Waste Management Paper No. 23* (DoE 1981). The UK Factory Inspectorate has produced booklets on toxic materials and how to measure their concentration in the atmosphere of a factory.

Acid iron wastes

Strongly acidic iron wastes result from the manufacture of titanium dioxide, which is used as a substitute for lead compounds in paints, varnishes, and lacquers and in the paper and plastics industries. Minor amounts are used in the manufacture of floor coverings, rubber, and printing inks. The main sources of titanium are the minerals ilmenite and rutile. The plentiful but low-grade mineral ilmenite (50–60 per cent TiO_2) is extracted by a lengthy process which produces sulphurous fumes and large quantities of acid iron wastes. A plant in Italy is reported to produce 1.5 million tonnes of this waste a year which is dumped in the Tyrrhenian Sea between Italy and Corsica. Similar plants in USA have been dumping in the sea off the New Jersey coast.

The scarcer mineral rutile (95 per cent TiO_2) is roasted in an atmosphere of chlorine at 1000°C. Titanium tetrachloride is produced with other metal chlorides and separated by cooling. The resulting liquid titanium tetrachloride is first distilled to an impurity-free product and then oxidized to titanium dioxide. As the rutile contains fewer impurities, there is less waste from a chloride plant than from a sulphate plant.

The acid iron wastes can be used as a source of iron but the high acidity and high levels of vanadium (900 mg/kg), chromium (140 mg/kg), lead (100 mg/kg), and cadmium (0.03 mg/kg) make dumps and lagoons potentially dangerous so the wastes are usually dumped in the sea or old mineshafts. Iron wastes are also produced in the refining of aluminium from bauxite (see *Red mud*, this chapter).

Acidity

Most soil-forming processes tend to increase soil acidity, but other natural processes buffer the change so pH rarely falls below pH 4.5. However, industrial processes and other exhaust gases may produce acidity in excess of any natural situation, as in the case of the so-called 'acid rain' that appears to be having disastrous effects on the poorly buffered soils of Northern Europe.

Coal shale heaps and opencast coal sites may develop strong acidity as a result of their significant content of pyrites (FeS_2). On first exposure, the shales are neutral or only slightly acid, but weathering and oxidation of the pyrites produces strongly acid conditions. The pH value of colliery shale depends

Table 4.1. Standards adopted in the Netherlands for soil contaminants. A, reference value below which soils are probably uncontaminated; B, value above which there is need for further investigation; C, value above which a clean-up is indicated. (From Moen *et al.* 1986)

Substance	Concentration in					
	Soil (mg/kg dry weight)			Ground water (μg/l)		
	A	B	C	A	B	C
Metals						
Cr	100	250	800	20	50	200
Co	20	50	300	20	50	200
Ni	50	100	500	20	50	200
Cu	50	100	500	20	50	200
Zn	200	500	3000	50	200	800
As	20	30	50	10	30	100
Mo	10	40	200	5	20	100
Cd	1	5	20	1	2.5	10
Sn	20	50	300	10	30	150
Ba	200	400	2000	50	100	500
Hg	0.5	2	10	0.2	0.5	2
Pb	50	150	600	20	50	200
Inorganic pollutants						
NH (as N)	—	—	—	200	1000	3000
F (total)	200	400	2000	300	1200	4000
CN (total free)	1	10	100	5	30	100
CN (total complex)	5	50	500	10	50	200
S (total)	2	20	200	10	100	300
Br (total)	20	50	300	100	500	2000
PO (as P)	—	—	—	50	200	700
Aromatic compounds						
Benzene	0.01	0.5	5	0.2	1	5
Ethyl benzene	0.05	5	50	0.5	20	60
Toluene	0.05	3	30	0.5	15	50
Xylene	0.05	5	50	0.5	20	60
Phenols	0.02	1	10	0.5	15	50
Aromatics (total)	0.1	7	70	1	30	100
Polycyclic aromatic compounds (PCAs)						
Naphthalene	0.1	5	50	0.2	7	30
Anthracene	0.1	10	100	0.1	2	10
Phenanthrene	0.1	10	100	0.1	2	10
Fluoranthene	0.1	10	100	0.02	1	5
Pyrene	0.1	10	100	0.02	1	5
Benzo(a)pyrene	0.05	1	10	0.01	0.2	1
Total PCAs	1	20	200	0.2	10	40

Table 4.1. *Continued*

Substance	Concentration in					
	Soil (mg/kg dry weight)			Ground water (μg/l)		
	A	B	C	A	B	C
Chlorinated organic compounds						
Aliphatic chlor. comp. (indiv.)	0.1	5	50	1	10	50
Aliphatic chlor. comp. (total)	0.1	7	70	1	15	70
Chlorobenzenes (indiv.)	0.05	1	10	0.02	0.5	2
Chlorobenzenes (total)	0.05	2	20	0.02	1	5
Chlorophenols (indiv.)	0.01	0.5	5	0.01	0.3	1.5
				0.01	0.3	1.5
Chlorophenols (total)	0.01	1	10	0.01	0.5	2
Chlorinated PCA (total)	0.05	1	10	0.01	0.2	1
PCB (total)	0.05	1	10	0.01	0.2	1
EOCl (total)	0.1	8	80	1	15	70
Pesticides						
Organic chlorinated (indiv.)	0.1	0.5	5	0.05	0.2	1
Organic chlorinated (total)	0.1	1	10	0.1	0.5	2
Pesticides (total)	0.1	2	20	0.1	1	5
Other pollutants						
Tetrahydrofuran	0.1	4	40	0.5	20	60
Pyridine	0.1	2	20	0.5	10	30
Tetrahydrothiophene	0.1	5	50	0.5	20	60
Cyclohexanone	0.1	6	60	0.5	15	50
Styrene	0.1	5	50	0.5	20	60
Fuel	20	100	800	10	40	150
Mineral oil	100	1000	5000	20	200	600

on: duration of exposure; content of pyrites; the abundance of $Al(OH)_3$; the presence of neutralizing materials such as calcium and magnesium carbonates; and whether or not the shale has been burnt. Oxidation of the pyrites takes place as

$$2FeS_2 + 2H_2O + 7O_2 = 2Fe^{++} + 4H^+ + 4SO_4^{--}$$

followed by

$$4Fe^{++} + 10H_2O + O_2 = 4Fe(OH)_3 + 8H^+.$$

Strong acidity also brings manganese and aluminium into solutions which are toxic to some plants. Mine drainage waters may be acid for the same reasons and pollute surface streams by acidifying them or by the precipitation of iron oxides in them. While the equations describe the oxidation in purely chemical terms, it is known that bacteria assist. In acid conditions the bacterium *Thiobacillus ferroxidans* is capable of oxidizing pyrites to ferrous iron and sulphur:

(1) $2FeS_2 + 7O_2 + 2H_2O = 2FeSO_4 + 2H_2SO_4$
(2) $4FeSO_4 + O_2 + 2H_2SO_4 = 2Fe_2(SO_4)_3 + 2H_2O$
(3) $FeS_2 + Fe(SO_4)_3 = 3FeSO_4 + 2S$
(4) $2S + 3O_2 + 2H_2O = 2H_2SO_4$

The first and fourth of these equations can take place as an independent chemical action, but the second and third require the participation of bacteria. It has been suggested that these reactions take place in succession (stages 1–3) depositing a sulphur-rich layer on the pyrites, which is then protected until the sulphur is oxidized as indicated in the fourth

Table 4.2. Tentative trigger concentrations for selected inorganic contaminants

Contaminants	Planned uses	Trigger concentration (mg/kg air-dried soil)
Group A: Contaminants which may pose hazards to health		
Arsenic	Domestic gardens, allotments	10
	Parks, playing fields, open space	40
Cadmium	Domestic gardens, allotments	3
	Parks, playing fields, open space	15
Chromium (total)	Domestic gardens, allotments	600
	Parks, playing fields, open space	1000
Chromium (hexavalent)	All uses	25
Lead	Domestic gardens, allotments	500
	Parks, playing fields, open space	2000
Mercury	Domestic gardens, allotments	1
	Parks, playing fields, open space	20
Selenium	Domestic gardens, allotments	3
	Parks, playing fields, open space	6
Coal tar (excluding skin contact)	Domestic gardens, amenity areas	200
	Public open space	500
	Industrial (without landscaping)	5000
Phenols	Domestic gardens and all areas where associated with mains water services	5
	Other uses	100
Cyanide (free)	Domestic gardens, amenity areas, and open spaces	50
	Industrial (without landscaping)	500
Cyanides (complex)	Any areas where plants are to be grown	50
Thiocyanate	All uses	50
Sulphur	All uses	1000
Sulphide	All uses	250
Sulphate	Residential, commercial, and industrial developments without landscaping	1000
	Domestic gardens, open spaces, landscaped areas, and other uses	200
Group B: Phytotoxic contaminants which are not normally injurious to health		
Boron (water soluble)		3
Copper (available—extracted by 0.05 M EDTA)		50
Nickel (available—extracted by 0.05 M EDTA)		20
Zinc (available—extracted by 0.05 M EDTA)		130

Soluble hexavalent chromium extracted by 0.1 M HCl at 37°C; solution adjusted to pH 1.0 if alkaline substances are present.

In Group B a pH value of 6.5 is assumed; if pH falls the toxic effects and uptake of these phytotoxic contaminants will be increased.

Phytotoxic effects of copper nickel and zinc may be additive. The trigger values are those applicable to conditions met in acid sandy soils; in neutral or alkaline loamy soils phytotoxicity is unlikely at these concentrations.

(This table is Crown Copyright and is reproduced here by permission of the Department of the Environment, from ICRL 1983.)

equation. This may partly explain the extended period of acid release from shale materials.

The development of strong acidity in restored tips and opencast coal sites depends both on the total content of pyrites present and its grain size. Very finely divided pyrites is oxidized more quickly and easily. Bradshaw and Chadwick (1980) suggested that 'if the pyritic sulphur content is multiplied by the percentage of reactive pyrites, then something like the right prediction for acid production can be made'. Acidification starts at the surface and gradually spreads downwards through the material.

Alkaline and saline wastes

Alkaline or saline conditions on derelict sites can

also pose problems. Alkaline wastes were produced by the now obsolete Leblanc process in the manufacture of sodium carbonate. This involved heating sodium chloride with sulphuric acid and then heating the resulting sodium sulphate to high temperatures with coal and limestone in a reverberatory furnace. Sodium carbonate was dissolved out of the furnace product and crystallized as $Na_2CO_3 \cdot 10H_2O$. Hydrochloric acid and sulphur were saleable by-products, but lime wastes (calcium hydroxide and calcium sulphide) were usually dumped near the factory premises.

Sodium carbonate is also produced by the Solvay process in which brine is saturated with ammonia, filtered, and then allowed to trickle down a carbonating tower against a countercurrent of carbon dioxide

$$NH_3 + NaCl + H_2O + CO_2 = NaHCO_3 + NH_4Cl.$$

The sludge emerging at the bottom of the tower is filtered and then heated to give sodium carbonate and steam. The sodium carbonate can be heated with calcium hydroxide to produce caustic soda with calcium carbonate as a by-product which may be dumped.

The manufacture of sodium chromite for the tanning industry also produced extremely alkaline and phytotoxic wastes (see *Chromium*, this chapter).

Gypsum (calcium sulphate) is produced as a waste product of the manufacture of orthophosphoric acid and hydrofluoric acid from phosphate rock. In the UK, two million tons of phosphogypsum are discharged into the sea or the estuaries of the Severn or Trent each year. Some is lagooned and disposed of on land but the liquid leachate from the lagoons contains fluoride in sufficient quantities to pollute watercourses. Work is in progress to find a suitable commercial outlet for this material.

Salinity has been encountered on some restored opencast coal sites in the Durham and Northumberland coalfield, where some shale materials contain substantial quantities of sodium chloride, potassium chloride, magnesium sulphate, and calcium sulphate. The saturation extracts of soils from Roddymoor, Durham, had conductivities of between 1 and 25 mmhos/cm, sufficient to inhibit plant growth (Doubleday 1971).

Arsenic

Although well-known as a poison in fiction, arsenic is not nearly as dangerous as many other metallic substances. It has been used as a nerve tonic, and low doses of it have been observed to prolong the life of laboratory animals. Arsenical compounds are widely used as insecticides and herbicides, as the active component in sheep dips, as a decolorant of glass, and as a pigment in the textile industry.

The element arsenic is found in 'native' form, but more commonly as its sulphide ore, arsenopyrite, or mispickel (FeAsS). Elevated levels of arsenic can occur naturally in stream waters draining from areas rich in mineral veins containing the element. It is usually associated with sulphide ores of copper, lead, and zinc and virtually all the industrial demand for arsenic is met by the by-products from smelting these metals. Thus, arsenic may be a problem near old smelting works. Flue dusts and emanations from smelter chimneys were particularly rich in arsenic and may have polluted the surrounding area. The brickwork of the smelters also contains raised concentrations of arsenic. The arsenic-rich material may have been spread around the site or taken elsewhere when the buildings were demolished.

Most coal contains pyritic material, including some arsenopyrite, which releases arsenic into the atmosphere on burning. The 3.3 billion tons of coal burnt in the world in 1970 released an estimated 8250 tons of arsenic. In the UK a 'trigger' concentration of 10 mg/kg arsenic in air-dried soil is recommended for domestic gardens and allotments and 40 mg/kg for parks, playing fields, and open spaces (Table 4.2). In the USA, the threshold limiting value for a 40-hour week exposure in workroom air (TLV/40) is given as 12 $\mu g/m^3$ (American Conference of Government Industrial Hygienists 1974). According to the Dutch recommendations (Table 4.1), the presence of 30 $\mu g/l$ arsenic in ground- and surface-water suggests the need for further investigations and a cleansing operation would be necessary where values exceed 100 $\mu g/l$.

Asbestos

The chemical inertness, heat resistance, and mechanical strength of asbestos make it very suitable for thermal and acoustic insulation and fire protection so that it is found in a wide variety of industrial and domestic situations. Its use is now sharply contracting but the presence of asbestos on derelict sites is a strong possibility and surveyors should always be vigilant, especially where any thermal lagging might have been used.

There are four fibrous silicates to which the general name asbestos is given: **actinolite** (asbestos ss) is a calcium–magnesium–iron silicate; **chrysotile** (white asbestos) is a fibrous serpentine with the composition of a hydrous magnesium silicate; **amosite** (brown asbestos) is a fibrous amphibole, also a magnesium–iron silicate; and **crocidolite** (blue asbestos) is a fibrous soda-amphibole. Asbestos minerals are mined mainly in the USSR and Canada with several smaller producers such as South Africa and Zimbabwe.

The ill effects of asbestos come from inhaling the dust, but this may not become apparent until one or two decades after exposure. The problem was first recognized in the 1920s and the British government enacted regulations to limit the exposure of workers in the asbestos industry in 1931 (Asbestos Industry Regulations) and in 1969 (Asbestos Regulations). This legislation has been reinforced by the 1974 Health and Safety at Work Act. Meanwhile, the use of crocidolite (considered to be the most dangerous type) has ceased in Britain and there is pressure to tighten still further the regulations concerning the amount of asbestos fibres permitted in the atmosphere at places of work.

Asbestos has been widely used for lagging or insulating pipes and vessels in many chemical engineering processes and power stations. Most power stations built before 1969 used crocidolite for insulation. Some 40 coal- and oil-fired power stations in Britain are to be decommissioned during the next few years and each contains several hundred tons of asbestos.

The transport industry has used much asbestos too for heat and sound insulation. Railways and shipping both employed it, so any premises, where railway carriages were built or dismantled, or shipyards, where ships have been constructed, refitted, or dismantled, could be polluted by asbestos.

It may also occur in any building or premises formerly used for the manufacture of asbestos wool, string or tape, paper or gauze, fire blankets, and protective clothing, as well as items like beer mats and filters. Asbestos is also used in the manufacture of brake linings for motor vehicles and so will be a contaminant present in former brake manufacturers' premises. While it might be released from brake linings on use, the heat generated in braking may modify the structure of the asbestos into a harmless form.

Thus derelict or contaminated land may contain asbestos from the demolition of buildings, dismantling of industrial plant, or from its many applications as an insulating material. Many buildings included asbestos board, or asbestos sprayed as an insulating layer on walls or ceilings to prevent the spread of fire. Demolition or the disturbance of rubble from old buildings may produce dust containing asbestos fibres; personnel need prior warning of the danger, as protective masks must be worn. The legal position in the UK is that exposure to asbestos dust shall not exceed 0.2 fibres of crocidolite per cm^3 of air when measured over any 10-min period or 2 fibres of other types of asbestos when measurements are averaged over a 4-h period. Short-term exposure should not exceed 12 fibres per cm^3 of air when measured over any 10-min period (DoE 1977*a*).

Sixty-two thousand tonnes of asbestos waste were disposed of in Britain in 1976 mainly to landfill but at least half of this amount was in the form of bonded asbestos in cement boards or resin-impregnated materials from which there should be no danger (DoE 1978*c*). Even though asbestos should have been disposed of correctly in landfill, the surveyor should always look out for asbestos dumped on derelict sites. There is danger from both correctly disposed and fly-tipped asbestos if it is disturbed.

It is important that asbestos on a derelict site be identified and located in a site survey (Smith 1980). At all times the creation of dust containing asbestos should be avoided. Where it is impossible to avoid it, the dust should be controlled by enclosure and fibres trapped from the exhaust ventilation. Where there is a danger of asbestos, protective clothing should be worn, which should be washed under special arrangements and not taken home (Michaels and Chissick 1979; ICRCL 1985).

Buildings

Buildings may have occupied all or part of a derelict site; if still standing they may be in a dangerous condition. Theft of tiles or slates and lack of maintenance allows the weather to penetrate the structure and deterioration occurs. Demolition of these buildings can provide a source of valuable hardcore (see *Rubble*, this chapter). Reinforced concrete or ferro-concrete structures were employed in many industrial buildings from 1900. The combination of the tensile strength of steel and the compressive strength of concrete provides a structure which will cause problems and extra expense during demolition.

Below the ground surface, heavy industrial plant would have required substantial foundations which may remain buried after the site has been levelled. Concrete floor slabs may be broken by earth-moving machinery but foundations of tall chimneys and blast furnaces require breaking down with pneumatic drills. Underground tanks, vats, and pipework frequently contain noxious fluids which may have to be removed by specialist contractors. Mineshafts and old wells have to be infilled and capped for safety.

Cadmium

Cadmium is used in alloys for bearings, solders, for electroplating, and as a fusible alloy in fire extinguishers. Traditionally, cadmium salts have been used as pigments in glass manufacture and more recently in nickel–cadmium batteries, semiconductors and as a stabilizer for plastics. Cadmium is very resistant to corrosion by seawater and so has been used as a protective coating on metals exposed to the sea.

Cadmium occurs as the sulphide ore, greenockite (CdS), which is almost invariably found in association with the zinc ore, sphalerite. Industrially, it is almost entirely produced as a by-product from zinc smelting and is rarely mined separately. Any contamination resulting from mining and concentrating cadmium is coincidental with that of zinc mining.

Soils and vegetation surrounding any base-metal mining or smelting operation may also be contaminated with cadmium. Davies and Roberts (1978) suggested that cadmium contamination around mines and smelters in north-east Wales is not as extensive as lead contamination but the overall pattern was the same with an upper limit of 540 μg/g dry soil. The moss-bag surveys by Goodman *et al.* (1975) in South Wales showed a similar distribution.

In natural conditions cadmium is found normally in very low concentrations but anomalously high levels occur at Shipham in Somerset which do not appear to adversely affect the inhabitants (Barltrop 1986). Yost (1984) traced the path of cadmium through the human food chain and environment. Experimental work has shown that cadmium is actively taken up into lettuce and cabbage from polluted soils (Alloway and Morgan 1986; Ottevanger 1986).

The 'trigger concentration' (Table 4.2) suggested by ICRCL (1983) is 3 mg/kg Cd in air-dried soil in domestic gardens and allotments with 15 mg/kg in parks, playing fields, and open spaces. The Dutch figures (Table 4.1) suggest that 1.0 mg/kg is permissible as a background level, at 5 mg/kg there should be further investigation, and above 20 mg/kg requires a clean-up operation. Levels of 2.5 μg/l in surface- and ground-water should be investigated further and 10 μg/l should be dealt with expeditiously (Table 4.1). The concentration of cadmium in workroom air (TLV/40) should not exceed 5 $\mu g/m^3$. (Am. Conf. Govt. Indust. Hygienists 1974).

Chromium

Chromium salts have a long history of use as a pigment. More recently chromium has become important as an alloy in stainless steels, special steels for high-speed tools, and in chromium-plating. Chromium oxides are used in the manufacture of refractory bricks and chromium salts in leather-tanning.

The mineral crocosite or lead chromate ($PbCrO_4$) occurs in the Urals, Brazil, and the Philippines. In Africa, chromite, or chrome iron ore $(Fe,Mg)_2CrO$ is the major source of metallic chromium and is associated with basic or ultrabasic rocks. It is obtained from deep mines so only the immediate surroundings are likely to be contaminated during mining operations. The metal is extracted by smelting with sodium carbonate alone or in a mixture with calcium carbonate. In the former process, the chromite reacts to form sodium chromite which is subsequently oxidized to sodium chromate. The soluble chromate is leached out but considerable quantities of chromite remain in the waste as a potential source of pollution. The second process forms calcium chromate and calcium chromite, plus complexes of both salts, all of which are soluble in acid. The chromium salts are removed by leaching but the residue still contains sodium chromate and calcium salts as pollutants. Even after a second leaching process, the greyish-yellow waste products from the chromium smelters may contain up to 2 per cent water-soluble chromium (Gemmell 1973). They are dumped, but as they contain crystalline sodium sulphate which is gradually leached away, there may be some subsidence of the tip surface. There are also problems from tip drainage and dusts. According to Gemmell (1973), the environmental implications of tipping these residues are largely unknown, but it is known that chromates are highly soluble and toxic to plants, animals, and man. Currently, chromium is produced in an electric furnace which reduces the oxides with carbon.

Chromic salts as solids, liquids, or sludges may contaminate land formerly occupied by the manufacturers of dyestuffs, paint-works, and tanneries, and there may be up to 10 g/m^2 Cr in the wastes from coal carbonization plants for gas, coke, or tar (ICRCL

1979*a*). The sites of many former engineering works may also be contaminated. Chromium is also a waste from colour-printing; a water-based copper–chrome–arsenic wood preservative may be present in former timber yards and chromic salts may have been used on many former chemical works.

In the natural environment, chromium is present in very small amounts; igneous rocks contain up to 100 mg/kg, shales up to 90 mg/kg, sandstones 35 mg/kg, and limestones 11 mg/kg. Normal soils may contain up to 100 mg/kg, but the range can be considerable with the highest levels in soils developed from serpentine or basalt.

The tentative trigger concentration adopted in Britain (Table 4.2) is 600 mg/kg chromium in dry soil for areas designated for domestic gardens and allotments and 1000 mg/kg for parks, playing fields, and public open space. However, the trigger concentration for hexavalent forms of chromium (Table 4.2) is 25 mg/kg for all types of land use. The Dutch authorities (Table 4.1) accept 100 mg/kg for normal soils; more than 250 mg/kg requires further investigation, and more than 800 mg/kg requires a cleansing operation. In water, up to 20 μg/l is normal, 50 μg/l requires further investigation and over 200 μg/l needs a clean-up operation. 2.5 $\mu g/m^3$ in workroom air is the maximum TVL/40 allowed in the USA (Am. Conf. Govt. Indust. Hygienists 1974).

Coal tars

Coal tar is a residue produced in the carbonization of coal in tar, gas, or coke works. It is a very complex mixture of benzene, toluene, xylenes, ethyl benzene, styrene, phenol, cresols, xylenols, high-boiling-point tar acids, naphtha, naphthalenes, acenaphthalene, fluorene, diphenyl oxide, anthracene, phenanthrene, carbozole, tar bases, and medium soft pitch. These substances may be separated by further distillation and form the raw materials for many chemical syntheses, e.g. benzene is used as the starting point for the preparation of aniline, phenol, and picric acid. Various oils may be separated for lubricants and creosote is used to preserve wood. Many of the components of tarry wastes are carcinogenic.

The petroleum-refining industry uses concentrated caustic soda to draw off the sulphur compounds from petrol, kerosene, and oils; concentrated sulphuric acid is used to refine the 'cracked' crude oil further. Tarry residues remain from both processes containing polynuclear aromatic hydrocarbons, some of which are carcinogenic, and the spent caustic soda also contains phenols, inorganic sulphides, and thiocyanates. A smaller quantity of tarry wastes results from the pharmaceutical industry. An example of contamination by coal tars at Uithoorn, The Netherlands is given by Yland (1986). Their corrosivity, inflammability, odours, and possible problems from skin or inhalation contact make them a hazard for human beings.

DoE (1977*d*) recommends that tarry wastes not be mixed with domestic rubbish in landfill. Their strongly alkaline or strongly acid properties could upset bacterial activity in a landfill site.

The trigger concentrations recommended by ICRCL (1983) for coal tar (Table 4.2) are 200 mg/kg for areas with houses and gardens, 500 mg/kg where public open space is planned, but levels of 5000 mg/kg can be allowed where industrial areas without landscaping are planned. The trigger concentrations (Table 4.2) for phenols are 5 mg/kg for domestic gardens and sites where mains water pipes are buried, but the level may be allowed to rise to 100 mg/kg where other uses of the land are contemplated.

The Department of the Environment has yet to suggest any trigger concentrations for the individual constituents of coal tars. However, the Ministry of Housing, Planning, and Environment (Netherlands) (1983) has proposed reference values (Table 4.1) as follows: benzene 0.01, 0.5, 5.0 mg/kg; ethyl benzene 0.05, 5.0, 50.0 mg/kg; toluene 0.05, 3.0, 30.0 mg/kg; xylene 0.05, 5.0, 50.0 mg/kg; styrene 0.1, 5.0, 50.0 mg/kg; naphthalene 0.1, 5.0, 50.0 mg/kg; anthracene 0.1, 10.0, 100.0 mg/kg; and phenanthrene 0.1, 10.0, 100.0 mg/kg where the values are for safe, needs investigation, and needs cleaning up, respectively.

Colliery waste

The dominant waste materials on colliery sites are the shale, mudstone, siltstone, seat-earths, and sandstone (rarely limestone) brought up with the coal. When coal was dug by hand it was possible to separate the marketable coal from the 'pit dirt' at the coal face. Modern mechanical methods of extraction produce material approximately half of which is unsaleable. This has to be separated and discarded at the pithead or washery (Gutt *et al.* 1974).

Its disposal produces the tip heaps of shale waste characteristic of coalfield areas. Formerly, it was disposed of by the Maclean apparatus which gave tall conical tips, by overhead conveyor systems which gave high ridges, or by movable tramroads which gave embankments (Fig. 4.1). As the material was

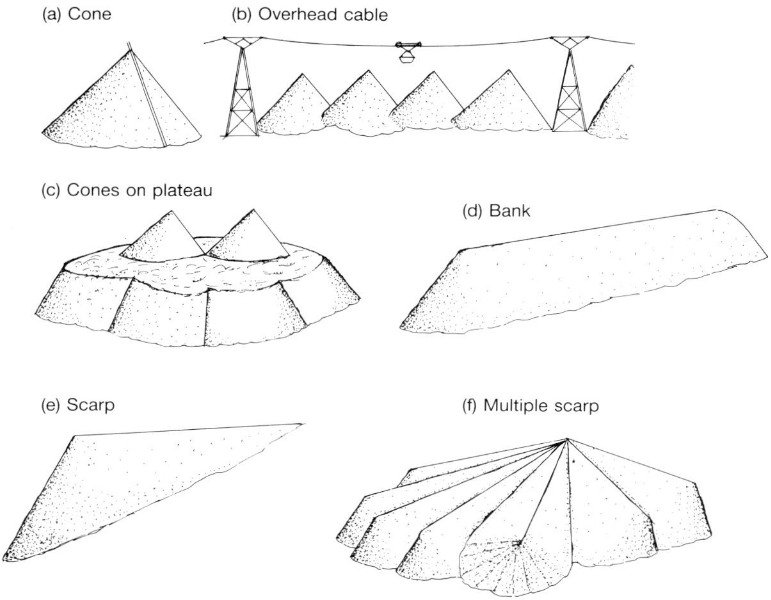

Fig. 4.1. Forms of colliery waste tips.

dumped by free-fall, it was loosely consolidated and permeable to air and water. Under heavy rain it could become saturated and debris slides occurred, sometimes with devastating results. Modern tipping practice constructs the heap in layers and compacts it by heavy earth-moving machines to prevent ingress of air and water. This greatly improves the stability of tips and has almost eliminated the risk of fire (Brent Jones 1983).

The mineral constituents of colliery waste consist principally of quartz, felspar, micas, clay minerals, haematite, magnetite, goethite, iron pyrites, and minor amounts of other minerals. The range of particle size can be considerable, including the larger fragments which have not been crushed and the silt-sized wastes from the washery which have been dried out in lagoons. The waste also includes ash, fragments of pit props, railway lines, sleepers, abandoned machinery, wire ropes, and hawsers, as well as fly-tipped rubbish and abandoned vehicles. In 1966 there were 1 184 678 000 tons of unburnt colliery shale on waste heaps owned by the National Coal Board (Gutt *et al.* 1974).

Many older colliery tips have been burnt. All coal shale tips contain substantial quantities of combustible material which can be ignited by the tipping of hot ashes, by surface fires, or possibly by spontaneous combustion. A burning tip is dangerous and emits fumes and dust that are a nuisance to the surrounding community. Burning greatly reduces the amount of combustible material, clay minerals are partly decomposed, carbonates completely decomposed, and iron sulphides may be changed to insoluble oxides, and there are some losses through volatilization. Burnt shale is reddish-brown and is generally a more stable material. In 1966 the NCB identified 628 620 000 tons of burning waste and 316 643 000 tons of burnt-out coal shale waste (Gutt *et al.* 1974).

In the 1970s, between 45 and 50 million tons of coal shale waste were being produced each year of which about 7 million tons were being used mainly for infilling in such projects as motorways. Burnt shale is preferred for fill and aggregate as it is more stable and less likely to ignite. Burnt and unburnt colliery shale can be used to replace some of the clay required in cement manufacture, brick-making, and aggregate manufacture (concrete blocks).

Although coal shale waste is not hazardous, the pyrites in it give a strongly acidic weathering product and some shales release soluble salts on weathering (see *Acidity* and *Alkaline and saline wastes*, this chapter). The surface layers of older tips have mostly completed such changes but regrading may reveal toxic materials beneath. Consequently, disturbance of old tipped material should be kept to a minimum and weathered material should be left on the surface. There may be strongly acid drainage waters from wastes rich in pyrites, but soluble salts are usually diluted to harmless levels on leaching.

On older, loosely compacted tips, slight changes such as removal of material from the toe of a tip or saturation by prolonged rainfall have been sufficient to induce rapid flow or slip. There are stringent safety requirements for tipping of colliery wastes and maintenance of the tip heaps. The water-table in tipped material can be monitored with piezometer tubes. Many older tips have been reshaped to conform with current safety criteria, contained in the Mines and Quarries (Tips) Act of 1969. Nevertheless, it is essential that any modifications of coal shale tips by earth-moving apparatus should not leave the tips in a critical condition.

When tips are to be removed, material should always be taken from the top of the tip and, where a burning tip has to be removed, breathing apparatus should always be available as the fumes and fine hot dust can cause asphyxiation. Water should not be allowed to seep into tips and increase their weight and instability. The use of water to reduce dust and heat could result in the production of water-gas and the danger of an explosion. Men and machinery can fall into voids caused by underground fires; if the fire is burnt out these voids may be dangerous but if still burning, they are lethal.

The character of domestic waste has changed considerably. Formerly it contained much ash and cinder from coal fires, but now paper, wrapping materials, and plastics provide the greatest bulk, and it is now more combustible. In addition, tarry and oily residues from chemical and engineering processes or paint solvents and old tyres may be added to the waste. With the included putrescent material, burning domestic refuse produces most obnoxious smells and the authorities try to keep burning to a minimum.

After studying underground fires, Beever (1986) found that there were many possible sources of ignition including surface fires, dumping of burning material, electrical faults, and spontaneous combustion. Laboratory experiments showed that smouldering combustion began at 193°C with woodflour and as low as 166°C for landfill site materials.

The calorific value of the materials gives some measure of their combustibility, but there are few guidelines established. A typical loamy soil has a calorific value of about 1.7 MJ/kg and it has been suggested that below a figure of 2.0 MJ/kg combustion is unlikely and that materials with calorific values above 10 MJ/kg are almost certainly capable of sustaining smouldering (ICRCL 1984).

Combustible materials

The presence of inflammable materials on a site can pose problems. The problem of combustible colliery shale has been dealt with in the previous section (*Colliery waste*), but many other materials found on derelict land and former landfill sites can also burn when ignited.

Materials on the sites of former gasworks, coke works, tar works, and oil refineries may be saturated in combustible liquids or tarry residues which give off unpleasant, if not dangerous, fumes if ignited. The iron oxides used for purification of town gas often contained up to 50 or 60 per cent of sulphur when they were finally disposed of, and there has been at least one case of sulphur dioxide fumes from a fire on gasworks residues (ICRCL 1979*a*).

Land which has been used to stock coal or coke, or for woodworking may be expected to contain potentially combustible residues. Railway land is particularly likely to be combustible as ashes and cinder were often used as fill when land was levelled for sidings. Some of these still contain sufficient unburnt coal dust and partly burnt fuel to be combustible. Docks and canal wharves may be similarly affected.

Copper

Copper has been used since prehistoric times. Its greatest uses now are in the electrical industry and as an alloy in the manufacture of brass and bronze articles, bells, and bearings for machinery. The salts of copper are used in the chemical industry as disinfectants, fungicides, and in the printing and dyeing of textiles. Copper contamination is most likely where the metal has been mined or refined.

Copper deposits are widely distributed as mineral-enriched veins, mainly in Palaeozoic rocks. The ore from these veins is high-grade and is usually won by a shaft mine and there may be waste heaps round the mine site. Production from these ores has declined steadily in favour of low-grade ores and most abandoned mines have been derelict for many years. Deep mines are still used to obtain copper ore from the Permian Kupferscheifer deposits of Germany and Poland and from similar but older sedimentary deposits in Canada, Michigan, Kazakhstan, and from the Zambia–Katanga copper belt in Africa (Skinner 1969).

In this century, copper production has become increasingly dependent upon porphyry copper ores.

These are low-grade deposits of chalcopyrites ($CuFeS_2$) with less than 2 per cent copper. They are mined from large, terraced, bowl-shaped open pits as at Bingham, Utah or Morenci, Arizona, both of which pits are over a mile in diameter and more than 400 m deep.

Whether the derelict land to be reclaimed has resulted from older mining activities or from modern open pits, there will be some contamination of the surrounding areas through downslope seepage of drainage water carrying copper, or the downwind spread of dust blown from the tips of waste material and particularly from dried-out tailings lagoons. In the process of ore concentration, reagents such as cyanide and hydrogen sulphide are added to the flotation tanks to aid mineral separation. Some of these compounds are biodegradable but others are toxic and may persist when mining has ceased. Baker and Bhappu (1974) suggested that these flotation reagents are used in minimal quantities and should be consumed completely in the process. However, plant failures or spills will have brought at least some of such compounds to the tailings ponds.

Metallic copper is obtained from high-grade ores by smelting in a blast furnace. The charge is mixed with coal or coke and molten copper is drawn from the bottom of the furnace. Some slag is produced and in Britain it is all sold for grit blasting (Gutt *et al.* 1974). Sulphide ores require a longer and more complex process which results in a greater amount of slag and gaseous pollutants. The ores are first heated in a reverberatory furnace with limestone or silica where the copper separates as a sulphide matte, part of the iron combines with silica to form slag, and a small proportion of the sulphur in the charge is oxidized. The liquid matte is then blown with air in a converter to remove all of the sulphur as SO_2 and the iron as oxides in a slag, to produce blister copper, containing about 0.5 per cent sulphur and 0.1 per cent oxygen. The blister copper is remelted and reduced to metallic copper by bubbling it with $CO+H_2$ or stirring with wooden poles. In modern processes the blister copper is purified by electrolysis which has the advantage that gold and silver may be separated as a by-product. Although this final process is relatively clean, the preceding concentration is less clean.

Dust and fumes discharged from the smelters will contaminate the soils and vegetation around copper smelters. Normal soils might be expected to contain 20 mg/kg total and 1 mg/kg extractable copper but analyses of soil up to 1 km from the smelters in the Lower Swansea Valley gave 900 mg/kg total and 40 mg/kg extractable copper. Throughout industrial South Wales the 'moss-bag' technique has revealed marked 'hot spots' in areas of former metallurgical activity (Welsh Office 1975).

Copper tips weather to a grey colour but black and reddish colours may be seen where they are disturbed. The texture of the tipped material is mainly coarse with some fines, but where slag has been tipped molten it will have cooled and fused into large porous masses. Reworking the slag for hardcore leaves the fused masses lying upon a very uneven topography. Even when undisturbed, copper tips are poorly vegetated.

A site where copper has been smelted or refined may also be contaminated by cadmium, mercury, nickel, lead, antimony, selenium, and zinc from the slag dust or in sludges from electrolytic methods of refining. Engineering works are often contaminated with copper especially where bronze castings and bearings have been manufactured; copper will also occur in metal wastes.

Works where wood has been treated with a copper–chrome–arsenic preservative and factory sites where copper-containing pesticides were manufactured are likely to contain raised copper levels. Copper is also used in the printing industry and in the textile industry as a dyestuff.

It is recommended (Table 4.2) that further investigations of a site take place when values exceed 50 mg/kg soil, or a zinc equivalent of >280 mg/kg in gardens and allotments. In water, Dutch recommendations (Table 4.1) are that a copper level above 50 μg/l should be investigated and if the concentration exceeds 200 μg/l there should be a clean-up operation. A TLV/40 of below 25 $\mu g/m^3$ in workroom air is recommended in the USA (Am. Conf. Govt. Indust. Hygienists 1974).

Cyanide compounds

The salts of hydrocyanic (prussic) acid are highly poisonous. Potassium cyanide is used in numerous chemical processes; sodium cyanide is used as a solvent in the recovery of gold and as a reagent in the separation of other metals such as copper from their ores. Limited amounts of cyanides may pass through the processing to the tailings ponds where they may contribute to toxic dusts. Sodium cyanide is also used in heat treatment processes in metallurgical plants and residues may contain up to 8 per cent of the active ingredient. Leakage and spillages on the site are distinct possibilities.

Manufacture of town gas at coal carbonization plants resulted in a product from which cyanide and hydrogen sulphide had to be removed by passing it through beds of iron oxide. The iron sulphide and ferric-ferrocyanide produced gave the spent oxides a dark-blue colour. This 'blue billy' often occurs on former gasworks sites. Thiocyanates and ferricyanide are also present in the ammoniacal liquor and the tarry residues produced, and may have contaminated soil and subsoil through dumping or spills during the life of the works. Cyanides are also used in the textile and printing industries as dyestuffs and colorants, and there has been limited contamination from these sources. Cyanides are a 'special waste' and must be correctly disposed of at controlled waste disposal sites. However, cyanide appears to be degraded in landfills and at one site where 4 tons were deposited only 1.5 per cent identifiable cyanide residues remained after ten years. Some of the cyanide had been converted to thiocyanate, with sulphur derived from the degradation of domestic waste, some was converted into gaseous hydrogen cyanide which escaped into the atmosphere, and the remainder reacted with metals to form insoluble precipitates such as cyanoferrate (II). Hydrogen cyanide gas has also been evolved where heat treatment chemicals have been consigned to landfill. Co-disposal appears to be effective, but decomposition of the cyanides by bacteria will be enhanced if the landfill is aerobic (Sumner Report 1978). Sodium cyanide oxidizes readily to carbonate, but the iron compounds are more stable.

The UK trigger concentration (Table 4.2) for cyanide in the soils of domestic gardens, allotments, amenity areas, and public open space is 50 mg/kg but 500 mg/kg is acceptable for industrial areas and car parks where the ground is covered by concrete or tarmacadam (ICRCL 1983). In the Netherlands (Table 4.1) a site with 10 mg/kg or more would need further investigation, and > 100 mg/kg would suggest an immediate cleaning operation; for complex cyanide compounds the corresponding figures are 50 mg/kg and > 500 mg/kg. In water over 50 μg/l of complex cyanide requires investigation and over 200 μg/l should be cleaned up without delay (Table 4.1).

Dioxins

This group of chemicals includes the most poisonous synthetic chemicals known. Dioxins occur as contaminants in the manufacture of several organic chemicals, particularly pesticides, and a number of accidents have drawn public attention to their dangers.

In 1976, 1800 ha of land round Seveso, Italy were contaminated with dioxins following an accident at a chemical works: 736 people had to be evacuated from their homes, 236 of them permanently; 3300 animals died and a further 77 000 had to be destroyed. Following attempts to cleanse the site 2 tonnes of contaminated soil in 41 barrels disappeared and were eventually found in a disused abbatoir in northern France (Tucker 1983). In this case £80 million has already been paid in compensation and a further 60 law suits remain to be settled.

An explosion at a chemical works near Bolsover in Derbyshire released dioxin. One man died and 79 developed skin conditions through contact with the dioxin. The contaminated equipment was eventually buried 60 ft deep at a site near Clay Cross, but it was necessary to hold a public enquiry to allay public fears about water pollution.

After being flooded in December 1982, the township of Times Beach on the Meremac river in Missouri was found to be heavily contaminated with dioxin. The contamination originated from waste oils sprayed on the unsealed roads to control dust. This is normal procedure but in this case the waste oil came from a chemical plant which had manufactured the herbicide 'agent orange'. Times Beach had the doubtful privilege of being the first township to be purchased by the Federal Government (for $33 m.) because it was considered too dangerous a place in which to live (Egginton 1983*a*, *b*).

Manufacture of weed-killers and insecticides at a Hamburg factory has produced 'more dioxin present than we could have imagined in our worst possible dreams'. The factory has closed as its management cannot reach the required environmental safety requirements (*Observer* 1984).

Problems arise in incineration plants where chemical residues are destroyed at insufficiently high temperatures to break down dioxins. Indeed, low-temperature incineration may be responsible for the production of these compounds from other organic molecules, especially in the presence of chlorine.

As yet no trigger concentrations for dioxins have been published, but it is claimed dioxin remains in human bodies for much longer than in animals and 2,3,7,8 tetrachlorodioxin is present in most people's fatty tissue. The daily intake should not exceed 0.1×10^{-12} g, and the only really safe figure is nil

(Mackenzie 1985). It should be assumed that sites of former pesticide and herbicide factories may be contaminated and they should be examined carefully.

Domestic refuse

Domestic wastes have traditionally been consigned to landfill, usually in old quarries and frequently on derelict sites or low-lying areas where it was desired to raise the level of the land. Old sites that received waste with considerable quantities of ashes will probably have completed settlement and will not present too many problems. The combustibility of recent domestic wastes can present problems with which the derelict land surveyor should be cognizant. Domestic waste now contains paper and timber as well as other organic matter and significant quantities of toxic elements, including lead, cadmium, mercury, zinc, copper, and nickel where co-disposal of toxic and non-toxic wastes has taken place (Gutt *et al.* 1974). Metallic elements are concentrated in ashes and clinker, both of which are consigned to landfill, and roadside drain sludge contains elevated levels of lead. Asbestos, sulphates, sulphides, and chlorides as well as phenols, cyanides, and fluorides may be present.

Microbial decay of organic matter in an anaerobic atmosphere produces gas consisting of up to 70 per cent methane with carbon dioxide, hydrogen sulphide, hydrogen, and nitrogen. Although decay may be complete in five to six years, the presence of toxic substances can retard microbial action and prolong it for 30–40 years. Disturbance of the site may rejuvenate the decay process and lead to further gaseous emisions. As these gases are asphyxiants and, in the case of methane and sulphuretted hydrogen, also explosive, problems may be expected (ICRCL 1978*b*).

Although landfill gases have been tapped as a source of power, they can be a hazard in site redevelopment. Seepage of gas into unventilated parts of buildings can lead to potentially explosive situations and it may be necessary to construct cut-off trenches with porous fill to allow gases to escape from the vicinity of buildings.

A common constituent of domestic waste, glass, can become a hazard on a site to be reclaimed for sportsfields. This problem can be solved by providing a sufficiently thick subsoil and soil cover to effectively seal the waste below the playing surface.

Where landfill sites have been in receipt of hospital wastes and slaughterhouse and knacker's yard wastes, there is always the possibility of the presence of pathogenic organisms which could be dangerous if disturbed. Anthrax spores are known to persist for at least 30 years.

The drainage of waters containing noxious fluids emanating from landfill sites can pollute streams and aquifers in the adjacent region. Pits with an impermeable base can become water-filled and overflow, and leachates from pits with a permeable base can contaminate groundwater.

Smells from landfill sites are nauseous but when decomposition has finished they should be minimal. However, if a site is to be reworked the disturbance could result in further smells and plagues of vermin or insects. 'Residual' smells may be very difficult to eradicate if land fill sites are redeveloped; disturbance should be avoided (ICRCL 1978*b*).

Dust

Dust problems are associated with mining, refining, and demolition; burning wastes may also lead to emission of dust.

Opencast mining operators are required to keep haulage roads moist to minimize the problem during dry weather. Concentration of ores is achieved by grinding the ore finely; it is then separated and the waste or tailings are run into a lagoon where they dry out. In a windy environment, dust from dried lagoons can be very troublesome. This has occurred with the gold mining wastes around Johannesburg (Clausen 1976).

Dust may be released from the chimneys of smelters and furnaces and blown on to surrounding land, taking with it contamination. Surveys by Goodman and Roberts (1971) and Goodman *et al.* (1975) revealed widespread atmospheric contamination by metallic dusts throughout South Wales with 'hot spots' occurring at steel works and sites of former smelters where the source of dust was the remaining waste tips. Davies (1983) reported high levels of metals in dust in former mining villages in Clwyd where lead, zinc, and cadmium were present. Beavington (1975) claimed that enhanced levels of copper and zinc occur up to 3 km away from an active smelter in Australia.

Toxic dusts can be a problem when a factory is to be demolished, especially when the premises are situated within an urban area. It may be necessary to bring in a specialized contractor to deal with toxic dusts before demolition work begins to comply with conditions for health and safety at work.

Colliery waste tips frequently burn for long periods producing fine ash which blows about, especially when the tips are disturbed. Campbell (1983) estimated that 7.6 g/s/ha could result from wind erosion and where a scraper was working, 110 $\mu g/m^3$ dust could be measured 1.3 km downwind.

Flooded workings

Abandoned gravel pits, brick clay pits, and many other quarries are liable to fill with water. The presence of water is a considerable advantage as it gives the site an attraction otherwise lacking. When vegetation recolonizes the banks and wildlife returns, these sites can become nature reserves through a policy of *laissez-faire*.

During their working life quarries below the water-table must be pumped; consequently, the water-table over a wider area is lowered. Water levels gradually return to their former levels when pumping ceases. If, subsequently, it is decided to landscape the area, a sudden draw-down of the water should be avoided as the saturated banks of the former quarry may collapse. (A similar situation arises in the reclamation of disused docks, especially those where a constant level of water has been maintained for many years and maintenance in recent times has been minimal.)

A further problem occurs in enclosed stagnant bodies of water if organic-rich material is available. The lower layers of the water may become strongly de-oxygenated and putrefaction may occur even though the upper layers remain aerobic. Removal of the surface layers can result in the emission of hydrogen sulphide. Miller (1983) gave an example of a drained quarry which could be smelt over 1 km downwind.

Fluoride

Fluoride is a local contaminant around brickworks, aluminium smelters, and manufacturers of phosphatic fertilizers.

In brickworks fluorapatite in clays is broken down during firing leading to the production of hydrofluoric acid in the flue gases. Also the electrolytic cells used in the production of aluminium are charged with alumina and natural or synthetic cryolite (Na_3AlF_6), and release carbon dioxide and some fluorides.

Small quantities of fluoride in drinking water (about 1 mg/kg) have been found to have beneficial action in prevention of tooth decay but where the concentration exceeds 20 mg/kg it may lead to hardening of the bones and inflammation of the joints accompanied by a characteristic discoloration of the teeth. Similar symptoms of fluorosis are observed in cattle grazing exclusively in the vicinity of brickworks or aluminium smelters.

Gases

The anaerobic decomposition of organic matter leads to the emission of gases from many disturbed areas, especially landfill sites and sewage farms. Natural methane production takes place in swamps and marshes; if ignited it forms the 'will-o'-the-wisp' phenomenon. In landfill sites, anaerobic decomposition of putrescent materials produces methane together with carbon dioxide, hydrogen sulphide, hydrogen, nitrogen, ethylene, and mercaptans (Parry and Bell 1983; Barry 1986; Carpenter 1986).

Where methane is generated (and natural gas leaks occur) the normal physical, chemical, and biological equilibria in the soil are disturbed. The aerobic soil atmosphere is replaced and anaerobic soil organisms increase in number. Movement of gases through the soil or landfill is dependent upon the texture and structure of the material; gas moves rapidly through cracks and fissures but diffuses slowly through clay. Voids in the soil or in landfill allow an accumulation of methane and accompanying gases, so it is advisable to treat all cavities with caution and to enter them only with breathing apparatus. Methane can be oxidized by bacteria. Also, bacterial activity may increase levels of carbon dioxide and depress oxygen levels when methane is present. The effect of landfill gas on soils and crops is described by Pankhurst (1973) and Hewitt and McRae (1986).

Normally, gases escape into the atmosphere and are dispersed harmlessly. However, if the site is reclaimed and buildings erected, methane can be a potential hazard. The gas seeps into service ducts, drains, and up through settlement cracks. As landfill sites are almost certain to have settlement problems, cracks in buildings are inevitable. Any underfloor cavity should be ventilated as methane is asphyxiating and inflammable; between 5 and 15 per cent it can explode (Bromley 1980).

Although less common, hydrogen sulphide is a highly toxic, flammable gas which is given off from landfill, sewage farms, and gasworks sites. H_2S can be asphyxiating and, in mixtures with air between 4.3 and 45.5 per cent, ignition in a confined space may lead to an explosion.

Similarly, hydrogen may be generated and could be

the cause of explosions. Where combustion has taken place below ground, carbon monoxide may be present in significant quantities (ICRCL 1979*c*; ICRCL 1984).

Gaseous emission can be monitored from boreholes or, where it is desired to monitor gases over a period of time, a perforated sampling pipe may be driven into the ground. When the equilibrium of gas in the pipe has been attained, sampling can begin. Although laboratory analyses would be more accurate, many gases and vapours can be measured by handheld equipment to a degree of accuracy sufficient for the proper maintenance of safety.

Studies of methane emission in the London Docklands led to the establishment of an emission rate of 0.3 $l/min/m^2$ as the level beyond which it was necessary to design buildings with anti-methane precautions (Thomson and Aldridge 1983). Attention should be paid to changes in atmospheric pressure as these can increase the rate of emission. Wind strength and direction will also affect the concentration of gases emitted.

Halogenated hydrocarbon compounds

The DoE (1976*i*) estimated that between 6000 and 10 000 tonnes of dry-cleaning fluids, mainly perchlorethylene and trichlorethylene, and 900 tonnes of metal-cleansing fluids are disposed of annually in landfill sites. These halogenated organic compounds have anaesthetic properties, and skin contact may lead to dermatitis. They are volatile and, as they might inhibit the action of bacteria in filter beds, they cannot be disposed of in sewage. They appear to be strongly adsorbed on to cellulose and plastic wastes and mineral oils when dumped in landfill. Investigations of landfill sites indicate that these substances do not appear in the leachate below a landfill site. However, experience in the Netherlands (where water-table levels are high) indicates that these substances can contaminate the ground water (Loch *et al.* 1986).

No trigger concentrations have been suggested for them by DoE, but the Dutch authorities (Table 4.1) advise that <0.05 mg/kg total PCBs might be regarded as uncontaminated, 1 mg/kg suggests further investigation is necessary and that anything in excess of 10 mg/kg may require a cleansing operation. The maximum allowable concentration of trichloroethylene in workroom air is 100 mg/kg during an 8-h day. However, decomposition of trichloroethylene or carbon tetrachloride can give rise to phosgene for which the maximum allowable concentration is only 0.1 mg/kg for an 8-h day.

Iron and steel works slags

During iron smelting and steel production, non-metallic impurities are complexed with lime and silica and concentrated in slag. The metal and slag are separately tapped from the furnace, the metal into ingots and the slag to be discarded on to the waste heap or 'slag bank'. However, a high proportion of the steel slag and virtually all of the blast furnace slag produced during the last century have found useful outlets and have not accumulated alongside the works, so most tips will be from iron- and steel-making processes of an earlier period and the composition of their outer layers will have been changed by weathering (Thomas 1983).

In modern processes, slag is tapped from the furnace at temperatures between 1400 and 1500°C and forms a calcium-alumino-silicate as it cools. Older slag may contain the calcium-alumino-silicate mineral larnite and calcium sulphate.

Most blast furnace slags can be used as aggregate but they may show volumetric instability. The sulphates present can lead to the formation of sulpho-aluminates (e.g. ettringite) with a 20 per cent increase in volume. Also, there may be volume changes on the transformation of dicalcium silicates from the α to the γ form. Weathering removes sulphates from the surface layers of tips, but the formation of sulpho-aluminates can begin again if the surface material is mixed with material lower in the tip.

In 1971, 71 blast furnaces produced 9 million tonnes of slag a year in the United Kingdom. Nearly all this slag was used for industrial purposes (79 per cent as roadstone and 16 per cent as rail track ballast, concrete aggregate, and filtration medium), so it did not accumulate at the works. Since then, the number of blast furnaces and the amounts of slag produced have declined (Gutt *et al.* 1974).

There is usually less steel slag as most of the impurities have been removed in the blast furnace. Steel slags may be acid or basic. Older acid slags contained low lime and high silica content, but few of them remain. Basic steel slags have a low silica content and a higher lime content, with free lime being present. They contain dicalcium silicate, free lime, divalent metal oxides, and calcium alumino-ferrite in solid solution (Gutt *et al.* 1974). Four million tonnes of steel slag were produced in 1971; of this 2 million tonnes were recycled and the remainder used industrially.

The presence of free lime and unhydrated magnesia can lead to volume changes in steel slags also, and the presence of lime and sulphur can have adverse effects on concrete structures and locally may pollute drainage channels.

The slag banks remaining are coarsely bedded and at an angle to the horizontal as a result of tipping. Sampling can be difficult because of their variability and it is often difficult to measure their potential for volumetric expansion.

Lagoons

The separation of minerals by concentration often produces lagoons of finely ground tailings. The sediments in a lagoon are not uniform as the coarser particles drop out of suspension nearer to the inlet pipe. Initially they have a low density of about 1400 kg/m^3 (Miller 1983) but once the lagoon has ceased to be used, a thin crust forms on the surface and it dries extremely slowly. On further drying the crust tends to sag in the centre where rain-water collects, so the contents of a lagoon can remain semi-liquid for many years after it has ceased to receive sediments.

Lagoons are usually constructed by damming a valley or gulley or are free-standing with sides of clay moulded into banks by earth-moving equipment. The banks may be raised as the sediment deepens. Where a former lagoon has been covered by later tipping, the slurry may be under sufficient pressure from the superincumbent weight to break a weak place in the lagoon wall with a sudden discharge of slurry. Miller (1983) described a lagoon purposely breached which discharged 10 000 m^3 of water which flowed for 1 km and another loaded lagoon which ruptured and released 'vast quantities' of water for 4–5 days and continued to flow for 6 months.

Lead

Lead is used in the automobile industry for storage batteries and in antiknock additives in petrol. Formerly, it was used widely in domestic water systems but its solubility in acidic waters makes it unsuitable for this use. It has been used by builders as a roofing material, particularly for churches, and for guttering, pipework, and flashing in domestic buildings. Formerly, it was widely used in paint manufacture and in the printing industry; it is highly likely to be a contaminant of former scrap-yards.

Lead is mined as a sulphide ore, galena (PbS), and usually occurs in association with zinc, copper, silver, cadmium, gold, antimony, arsenic, and bismuth, as a geothermal addition that lines the surfaces of faults and joints. Low-grade deposits of lead, similar to the porphyry copper ores have not been discovered. Consequently, lead is usually obtained from deep mines rather than opencast or open-pit mines and mining waste is limited to the immediate vicinity of the mine. In north Derbyshire, miners excavating lead from 'lead rakes' along lines of geological faults produced characteristic linear areas of disturbed ground, often now being reworked for fluorspar. Lead has also been obtained from the Carboniferous Limestone in the Mendip Hills and Halkyn Mountain in north-east Wales. In the Mississippi Valley, lead ores occur as replacement bodies in limestones, mainly of Palaeozoic age, probably derived from relatively low-temperature brines carrying metallic elements, such as have been observed in geologically active areas like southern California and the Red Sea rifts (Skinner 1969). In Germany and Poland, lead sulphides accumulated in the deeper parts of a structural basin to form the Kupferschiefer beds which are mined by deep mines. Lead is also obtained (with zinc) from the Mount Isa mine in Queensland from Pre-Cambrian sedimentary strata.

Lead ores are concentrated by crushing and flotation similar to those used for copper and zinc. This gives a concentrate with up to 75 per cent lead, because the high density of lead makes separation easier. In the past, as at the long-disused mines of the Ystwyth valley of Wales, rock waste and tailings were simply dumped on the valley side and in the river. Fluvial action has spread the toxic material downstream to pollute the alluvial soils and vegetation, affecting grazing livestock (Davies and Lewin 1974; Davies 1977). In Britain, the sites of derelict lead mines are now indicated by the heaps of rock waste. These are mainly fragments larger than 5 cm which form a droughty, poor substrate for plant growth. Tailings from the concentration process are finer, < 2 mm, and are usually found in lagoons or dried-out lagoons. They provide a better substrate for plant growth, but like the mine waste still contain about 1 per cent lead which represents a heavy contamination.

In the eighteenth and nineteenth centuries, lead ore was smelted in a reverberatory furnace to obtain metallic lead. The floor of the furnace sloped down to a well where the molten lead collected. As with copper and zinc smelting, copious fumes of sulphur dioxide escaped into the atmosphere. A second heating with calcium carbonate released the remaining sulphur dioxide and left almost pure lead. The remaining impurities were removed by skimming from the surface of the smelt during several days heating. This slag was either recycled or discarded with other waste material from the smelters. However, the smelters rarely produced lead alone, but they turned to whichever metal attracted the best market price. The slags which resulted were very mixed, but may be expected to include cadmium, mercury, nickel, antimony, selenium, as well as some copper, zinc, and lead. Lead is now smelted in a blast furnace and this process results in 60 000–70 000 tonnes of granular slag a year at the Avonmouth works of Imperial Smelting Processes, the bulk of which is being used to raise the level of land in the vicinity of the works where an industrial estate is being developed. Some is sold as aggregate and its use, for shot-blasting grit, bricks, as a water-filtration medium, and in night storage heaters, has been explored (Gutt *et al.* 1971).

As with copper and zinc smelting, the fumes of sulphur dioxide blighted the vegetation around the site of smelters. Davies and Roberts (1978) demonstrated residual contamination of soils in areas where lead was mined or smelted. They produced maps and three-dimensional diagrams based on a grid of 260 sampling points, which graphically illustrate the peaks in the geochemical landscape associated with mines and smelters. Altogether 66 per cent of the area surveyed had lead levels above 70 $\mu g/g^{-1}$ soil and was judged to be contaminated (Davies 1983). Further studies of the same area by Page and Swires-Hennessy (1986) revealed that air and water lead levels were satisfactory, but amounts in soils and dusts were above accepted safety levels. However, mean blood-levels did not exceed critical concentrations. Bridges *et al.* (1981) showed that extractable lead preferentially accumulated in the litter and A horizons of soils beneath woodlands on contaminated soils in the Lower Swansea Valley.

Lead must be expected as a contaminant on sites formerly occupied by engineering works. It is also found on most garage premises, and is increasingly evident in soils from roadside verges (Haines 1981). Scrap-yards are almost certain to contain elevated levels of lead, especially those which specialize in scrap batteries. Davies (1978) reported levels of lead eight times the normal background in a garden soil 200 m from a scrap battery works in the Manchester area. Former lead-based paints caused widespread contamination of urban areas where residues from burnt-off paints, discarded when houses were repainted or modernized, remain in garden soils. These paints have now been superseded by less toxic materials.

Lead is a likely contaminant in landfill sites and sewage works sites. The extent of contamination in landfill will be extremely variable and difficult to assess unless it is known whether lead-containing wastes have been dumped. Lead is likely to be a contaminant of sewage farms; lead levels in the soils of the Beaumont Leys site, Leicester, used for disposal of sludge and liquid effluent from 1890–1964, ranged between 47 and 2470 mg/kg soil (total) and 1 to 42 mg/kg soil (available) where sludge had been spread, and between 7 and 870 mg/kg soil (total) and 0.5 to 50 mg/kg soil (available) where the effluent was irrigated (personal communication, M. A. Smith).

In the environment generally the concentrations of lead present are approximately: igneous rocks 12.5 mg/kg; shales 20 mg/kg; sandstones 7 mg/kg; limestones 9 mg/kg. In soils lead is strongly associated with organic matter, so 10 mg/kg is an average topsoil value. However, 47 per cent of samples from 476 sampling points in the urban areas of Walsall exceeded a level of 100 mg/kg and 1.5 per cent exceeded 1000 mg/kg lead (JURUE 1982).

The trigger concentration (Table 4.2) for total lead content in Britain is 500 mg/kg for garden and allotment soils but up to 2000 mg/kg may be allowed where the land use is to be parks, playing fields, or public open space. The Dutch (Table 4.1) recommend rather lower figures; at 150 mg/kg further investigations should take place and at 600 mg/kg a clean-up operation is recommended. For waters the corresponding figures (Table 4.1) are 50 μg/l and 200 mg/kg, respectively. A TLV/40 value of 3.8 $\mu g/m^3$ is recommended in the USA (Am. Conf. Govt. Indust. Hygienists 1974).

Mercury compounds

Soluble mercury compounds are very poisonous and the liquid metal itself is sufficiently volatile to be highly toxic. Mercury has many industrial uses and may be a potential contaminant at many kinds of site.

Mercury is obtained from the sulphide ore, cinnabar (HgS), which occurs erratically in narrow hydrothermal veins. It is mined in Spain, California, Chile, and Queensland but cinnabar occurs as a contaminant in other sulphide ores such as copper, lead, and zinc. Mercury contamination will be associated with the smelting of these metals. Mercury is used extensively in the refining of gold and silver and a tin amalgam is used to back mirrors. Amalgams used to be used in dentistry.

Mercury occurs in sludges from the chlor-alkali industry; mercuric chloride is a major constituent of dry batteries and is also used in pharmaceutical products; mercuric sulphide is the pigment vermilion; and mercuric fulminate is sensitive to shock and is used as a detonator for explosives. Mercuric nitrate was used in the hat-making industry where mercuric poisoning formerly afflicted workers with 'hatters' madness'. Organic mercury compounds are used as fungicides.

Mercury is extensively used in the electrical and instrument-making industries; for example it is used for rectifiers, automatic switches, mercury vapour lamps, and in the familiar barometer or thermometer.

Apart from all these industries, it is inevitable that mercuric wastes are found on landfill sites. Reject and used batteries may contribute substantial quantities of mercury to landfill sites. However, a DoE survey (Sumner Report 1978) did not find any organo-mercury compounds in the lower levels of landfill because reducing conditions had converted them to mercuric sulphide. Mercury compounds appear to be strongly adsorbed on to montmorillonite clays, so this may be a means of encapsulating and isolating them.

The concentration of mercury in soils normally lies between 0.01 and 0.30 mg/kg and this generally decreases with depth. The recommended trigger concentrations (Table 4.2) for mercury are 1 mg/kg in soils for houses, domestic gardens, and allotments and 20 mg/kg for parks, playing fields, and open spaces. In waters, the Dutch (Table 4.1) recommend further investigation above 0.5 μg/l and the necessity for a cleansing operation above 2.0 μg/l.

Metal wastes

Wastes from a wide variety of metal-working processing including cleaning and pre-treatment processes, electroplating, anodizing, galvanizing, rustproofing, machining, and polishing may be disposed of in landfill sites. Leaching of metals from landfill sites may affect surface and ground waters. These wastes may be solutions, sludges, or solids (DoE 1976*h*). They may contain various compounds of copper, chromium, lead, zinc, nickel, cadmium, mercury, iron, or aluminium, together with noxious fluids associated with cleansing or pre-treatment. The major hazard of such deposits is from the leachate. However, it has been shown that many metal contaminants are effectively retained in the landfill, but if they escape as leachate they are strongly adsorbed in the rock immediately below the tip. Potentially harmful sludges were incorporated in an experimental landfill. They showed very little change over three years whether in aerobic or anaerobic conditions (Sumner Report 1978). If landfill sites occur above sensitive aquifers, it is possible to treat heavy metal sludges chemically to stop migration.

Discharge of waters from metal-processing works into rivers or aquifers is of great concern as it may cause pollution of drinking water supplies. Table 4.3 presents the major sources of metal-containing waste waters and their concentrations.

Nickel

Although nickel has only been widely used during this century, it is a minor constituent of the ores of copper, lead and zinc. Any district where these metals were refined in the eighteenth and nineteenth centuries now has an enhanced background level of nickel. The extent of land made derelict solely through the mining, refining, or use of nickel is very small and is mainly caused by sulphurous fumes from the first stages of refining. Nickel is used now as an alloy in the production of stainless steels, and steels which are to be subjected to high temperatures and high-speed working. Its most common use is in coins.

Nearly all the world production of nickel comes from the mineral pentlandite, mined in the Sudbury district of Ontario, Canada. This ore was formed by separation of nickel, copper, and iron sulphides from a molten mafic magma. Nickel is also concentrated by intense weathering of ultrabasic rocks such as peridotite, when nickel ions in the lattice of olivine are released and then redeposited as a nickel–silicate, garnierite, or as a nickel–iron sulphide, pentlandite. Nickel-rich lateritic deposits were first recognized in New Caledonia, but similar deposits are known to occur in Cuba, the Philippines, Borneo, Greece, and in other countries. The Sudbury sulphide ores are mined from deep mines, but lateritic ores would be extracted from wide, shallow, open pits.

In the early days, the ore at Sudbury used to be hand-sorted and then up to 100 tons were piled upon a foundation of wood. The wood ignited the sulphur in the ore and it continued to burn for three or four months, emitting sulphurous smoke. Ore concentration and the smelting of copper and nickel at Sudbury

Table 4.3. Major industrial sources and concentrations of metals contained in waste waters (Patterson 1979)

Contaminant	Sources	Concentration range (mg/l)
Arsenic	Insecticide manufacturing; ore extraction; sulphuric acid and ammonia manufacturing; wood products preservation; arsenic trioxide manufacturing	10–1000 (soluble > 100 mg/l)
Cadmium	Metallurgical alloying; ceramic and inorganic pigments; electroplating; textile printing; batteries; acid mine drainage	4–1000
Chromium	Corrosion control additives; inks; pigments and dyes; tanning; aluminium anodizing; metal cleaning, treatment, and plating; preservatives	1–1300 (wastes) (15–270 000 baths which may be batch dumped)
Copper	Metal cleaning, pickling and plating; rayon manufacturing; copper and brass manufacturing; copper sulphate production; primary and secondary copper manufacturing; electronics	10–900 (up to 7000 in primary and secondary copper manufacturing)
Iron	Mine drainage; coal storage pile drainage; steel manufacturing and rolling, inorganic chemicals; metal processing and plating; titanium dioxide, paint and ink manufacturing	40–5000 (pickle baths up to 100 000)
Lead	Battery manufacturing; electroplating; television tube manufacturing; glass and ammunition manufacturing; tetraethyl lead production; paint pigment; rubber hose manufacturing	< 1–400
Manganese	Mine drainage; refuse incinerator and industrial scrubber waters for ferromanganese and electrolytic manganese production; paint formulation	< 1–> 1000 (1000–6700 in one case)
Mercury	Chlor-alkali plants; paper mills, smelting; water-based paint; acetaldehyde production; fertilizer mills; chemical manufacturing	< 0.3–20 (older references 2000)
Nickel	Pickling and plating of nickel and nickel alloys; basic steel works, motor vehicles, and aircraft industries; business machine and tableware manufacturing	< 1–900
Selenium	Paint, pigment, and dye production; electronics and glass manufacturing; insecticide industry; pulp and paper production; non-ferrous metals smelting	< 1–10 (very limited data)
Silver	Jewellery and silverware manufacturing; metal alloying; food and beverage industry; photographics, ink formulation; electroplating	< 1–250
Zinc	Steel works (galvanizing); zinc and brass metal works; zinc and brass plating; groundwood pulp and newsprint paper production; inorganic pigments; electroplating and metal processing; zinc smelting; mine drainage	< 1–2000 (plating, pickle and bonderizing baths 2000–35 000)

has wrought considerable devastation around the mines and smelters.

The Chinese produced nickel–silver well before nickel was isolated elsewhere. Their earliest process was to dissolve the ore in sulphuric acid to separate the nickel and copper. Latterly, the ore has been concentrated by flotation methods and smelted in a reverberatory furnace to produce a copper–iron–nickel matte. The iron is removed in a converter to leave a copper–nickel matte. The copper and nickel are fused with sodium sulphate in the Orford process. The sodium and copper sulphides rise to the top and the nickel sulphide settles to the bottom of the furnace. After cooling, the bottom part is ground, and then sintered to the oxide which was reduced to metal by roasting with charcoal or flour. This method has been superseded by the Mond or electrolytic methods. The Mond method takes the nickel sulphide from the Orford process and heats it to 700°C in air to produce nickel oxide. The oxide is reduced to the metal by heating in an atmosphere of water-gas; carbon monoxide is then passed over the nickel to form volatile nickel carbonyl which separates the nickel from impurities. Nickel is deposited on pellets of nickel, heated to 180°C which releases the carbon monoxide for reuse. Alternatively, the product of the Orford process can be purified by electrolysing a nickel sulphate solution containing sodium sulphate and boric acid.

Earlier methods of smelting produced much sulphurous fumes and dust which fell upon the surrounding area. The electrolytic and Mond methods of nickel refining produce much less nickel emission. The Mond process, particularly, utilizes the highly toxic gas nickel carbonyl so it must be completely enclosed; releases should only occur when there is a plant failure.

As nickel is a contaminant of copper and zinc ores, smelters of these metals may have spread nickel in the surrounding district. An Atomic Energy Research Establishment survey showed a slight nickel contamination in the Lower Swansea Valley, and the 'moss-bag' survey of Goodman *et al.* (1975) clearly showed that these concentrations were centred upon metalliferous works. However, there appear to have been no reports of environmental damage by nickel compounds in the atmosphere or elsewhere, and nickel does not appear to have caused problems to vegetation, domestic animals, or wildlife.

Normal environmental levels of nickel are: igneous rocks 75 mg/kg; shales 68 mg/kg; sandstones 2 mg/kg; limestones 20 mg/kg; up to 40 mg/kg is normal in soils but higher levels do occur naturally (Skinner 1969). In the area between the former metalliferous works of the Lower Swansea Valley and the Mond nickel refinery at Clydach the nickel levels of soils and vegetation are well below commonly accepted danger levels (Goodman 1979). Sewage disposal sites also may show enhanced nickel levels. At the Beaumont Leys site at Leicester, available nickel levels were between 5 and 200 mg/kg where sludge had been incorporated into the soil and between 2.2 and 56 mg/kg where irrigated with effluent (Heeps and Pike 1980). The advisory service of the Ministry of Agriculture is concerned about nickel in sewage sludge disposed of on arable land and assumes that it is eight times more toxic than zinc. It recommends that a maximum of 30 years of sewage sludge application should not add more than 225 mg/kg of nickel to arable land or 450 mg/kg to pastures.

In Britain, further investigation is considered advisable if the trigger concentration (Table 4.2) of 20 mg/kg of available nickel is exceeded. The Dutch authorities (Table 4.1) recommend 100 mg/kg, and recommend that clean-up action should occur where levels rise above 500 mg/kg. The comparable figures (Table 4.1) for surface and ground waters are 50 μg/l and 200 μg/l. A TLV/40 of 25 μg/m^3 is recommended for workroom air in the USA (Am. Conf. Govt. Indust. Hygienists 1974).

Oil

As a result of the heavy dependence of modern society upon oil and oil products, petrochemicals are handled at many locations. Contamination by oil or petrochemicals may occur through spillage at refineries, leakage of storage tanks and pipelines, accidents with tankers on road and rail, or at docks. Coasts, and salt marshes in particular, have been polluted with oil from ship discharges.

Crude oil is a dark-brown odorous liquid containing paraffins, naphthalene, aromatic hydrocarbons, and hydrocarbons, together with smaller amounts of pyridine, quinoline, phenols, resins, mercaptans, and asphaltic compounds. The crude oil is distilled into heavy oils for heating, light fractions for petrol, diesel, kerosene, and many other by-products which find a ready outlet in plastics, pharmaceuticals, and other uses.

Heavy oil penetrates slowly into soil whereas the lighter fractions penetrate much more quickly. Thus it is practical to scrape up as much as possible of spills of heavy fuel oil and crude oil as quickly as possible before too much seeps into the soil. If the oil has percolated through the soil to pollute the ground water, boreholes can be sunk to the water table to draw off the oil floating on the surface. Clearing the oil by burning is not recommended (CONCAWE 1983).

After the worst has been scraped up and disposed of safely, it is probably best to allow the remaining oil to dry before cultivation. Soil aeration assists the escape of the volatile components but there may be structural damage if the subsoil remains saturated. Nitrogen and phosphate fertilizers are said to assist recovery by encouraging the soil microflora (Schwendinger 1968).

Contamination of soil with oil results in an immediate increase in its organic matter content. As bacteria can metabolize oil, they make a heavy demand upon soil oxygen which can result in the soil becoming anaerobic, particularly if the soil is blanketed with an oily layer which limits gaseous exchange. The effect of oil on soil structure is not clear; it has been claimed that it weakens soil structure, but other observations suggest that structure is improved. It is possible that the structure may be weakened initially, but later improved when the hydrocarbons are incorporated by bacteria (Plice 1948; Ellis and Adams 1961).

The effect of an oil spill upon plants depends upon the type and amount of oil. Crude oil contains more phytotoxic compounds than refined oils or petrol. Coastal marshes appear to recover from an oil spill after two or three years, and probably crops and grass also. Plants may suffer from toxic constituents in the oil, or blocking of their stomata, and the inhibition of their respiration. Lighter fractions may penetrate the epidermis to injure the internal structures. As a result, the yields of crops and grass are depressed by oil, with root crops being most affected. Tainted crops are unfit for human consumption (Baker 1971; Skinner 1973).

Recurrent spillages, as in an oilfield, a refinery, garage premises or a scrap-yard, could have penetrated deeply into the soil and spread lateraly from the point of contamination. Intense saturation raises some risk of combustion, and odours from the saturated earth will be difficult to eradicate. Complete removal and replacement of the soil may be necessary before the site is redeveloped. Huntjens *et al.* (1986) show that decomposition of oil compared with hexadecane ($C_{16}H_{34}$) is slow under laboratory conditions.

Oil forms a surface layer on ponds or streams, which fouls animals and birds and is unpleasant if it gets into water supplies; oily sludges may lie on the bottom of water bodies; lead compounds in petrol can be poisonous to animal life and can be passed up food chains to human beings. DoE (1976*g*) recommends disposal of oily wastes by incineration but inevitably some are placed in landfill sites. When this occurs, the Sumner Report (1978) concludes that oils may be adsorbed on to domestic refuse and, if quantities are kept low, there should be no oil in the drainage. Oily wastes may pose a risk of fire and contact with the human skin may cause irritation, dermatitis, and even cancer if sufficiently long exposure is maintained.

Oil shales and tar sands

There may be a future problem of reclaiming land contaminated by the mining of rocks rich in asphaltic hydrocarbons (oil shales and tar sands). There are an estimated 30 000 square miles of tar sands in northern Alberta, along the Athabasca river which contain the equivalent of 600×10^9 barrels of oil. Shales containing oil in Wyoming, Utah, and Colorado are thought to contain reserves of 850×10^9 barrels of oil. There are also oil-shale deposits in Estonia and Scotland but their exploitation is marginally economic while sufficient oil is available from conventional crude oil producers. To obtain oil from them the tar sands must be mined by opencast methods with considerable potential for creating derelict land.

Oil was first produced from oil shales in Scotland in 1865 and up to 1950 over 90 per cent of the world output of oil from shale came from Scotland; it is estimated that there are already 150 million tons of oil-shale waste in the Lothian region of Scotland. In the production process, the shales or sands are heated until the tarry extract flows out of the rock or is volatilized. It is then collected and refined into a range of petrochemical products in the same way as crude oil. The spent shales still contain some carbon and can be burnt as fuel at higher temperatures than are required for oil extraction. Spent shale is returned to the opencast mine or dumped on the surface. Between 80 and 90 per cent of the original weight of the shale remains but after processing its volume may be greater.

The particle-size range of the spent shales falls within the normal soil range but plant growth is limited because there may be difficulties in rewetting the material. It has a high pH, contains soluble salts, and a low level of plant nutrients, especially phosphorus and nitrogen. In the drier climate of the western USA the soluble salts must be leached and possibly gypsum added to displace the sodium ions. Once leached, and with a cover of topsoil, grass yields match those on undisturbed soils (Schmehl and McCaslin 1973; Davis 1978). In cooler and moister climates, Bradshaw and Chadwick (1980) reported that trees with low nutrient requirements grow successfully on restored sites in Estonia. Brown (1976) reported upon a feasibility study for the removal of the shale tips for land reclamation in the Firth of Forth.

In general a calorific value of above 2.0 MJ/kg indicates the possibility of combustion (ICRCL 1984). Dutch figures (Table 4.1) suggest that less than 100 mg/kg mineral oil in the soil presents little hazard but that more than 1000 mg/kg requires further investigation and more than 5000 mg/kg requires a cleansing operation. Several other compounds from oils might be found on sites where oil has been handled or dumped. Less than 0.1 mg/kg naphthalene can be considered uncontaminated, but further investigations should take place above 5 mg/kg and above 50 mg/kg it may be necessary to clean up the site (Table 4.1). The corresponding figures for benzene are 0.01 mg/kg, 0.5 mg/kg, and 5 mg/kg; for pyridine 0.1, 2 and 20 mg/kg; and for phenols 0.02, 1, and 10 mg/kg.

Paints

Contamination from paints arises from the constituents used for colouring, fillers, preservatives, and biocides as well as the various solvents used for dilution or cleaning. Pigments may have contained the salts of barium, chromium, cobalt, copper, lead, manganese, mercury, nickel, tin, titanium, and zinc. On sites where paint has been manufactured, scraped, burnt-off, or shot-blasted and added to the soil as fine particulate matter solvents may have been spilled (see *Solvents*, this chapter).

Pesticide compounds

The development and manufacture of pesticides (including insecticides, fungicides, and herbicides) may lead to dangerous contamination. The ICRCL committee (1979*b*) reported that advice has been sought on land contaminated with herbicides. DoE (1980) presented a general classification in Waste Management Paper 21, which follows (direct quote):

(a) Inorganic and organo-metallic pesticides including compounds which contain one or more of arsenic, boron, chromium, copper, fluorine, manganese, mercury, thallium, tin, and zinc, the metals of which remain after disposal. Other compounds include lime sulphur, ammonium sulphamate, aluminium phosphide, calcium cyanide, and sodium chlorate which either liberate toxic fumes or may form explosive mixtures.

(b) Phosphorus-containing pesticides used as insecticides in place of persistent organo-chlorine compounds.

(c) Nitrogen-containing pesticides, herbicides, and insecticides from the carbamates, amides, substituted ureas, triazines (simazine), and amines (DNOC and paraquat), which work by the inhibition of cholinesterase in the nervous systems of insects; most are also highly toxic to mammals. Paraquat is rapidly inactivated in the soil by adsorption on to soil particles.

(d) Halogen-containing pesticides are very persistent in the environment and are also toxic to man. This group includes the compounds DDT, aldrin, dieldrin, lindane, and heptachlor as well as the phenoxyacids, hormone weed-killers such as MCPA and 2,4,5 T. These substances can have unpleasant side-effects through the dioxins which they contain.

(e) Sulphur-containing pesticides such as parathion and thiocarbamates do not pose serious toxicity or environmental hazards.

(f) Biological pesticides include naturally occurring substances or closely related synthetic compounds. This group includes giberellic acid, pyrethroids, and strychnine.

All pesticides used in Britain undergo strict testing before being marketed (MAFF 1981). In conjuction with the Pesticides Registration Department at Harpenden, every product must undergo trials to assess its impact upon the target pest, upon workers handling the product, upon the food crops being protected, and upon the environment as a whole. Checking and monitoring are also carried out and the Pesticides Registration Department is responsible for ensuring the correct labelling and instructions given for use.

Many modern insecticides act upon the nervous system of insects by blocking the action of acetycholine so that the nervous system breaks down and the insect is killed. Natural organo-chlorines such as pyrethrins are effective insecticides but do not last long on exposure to air. Synthetic compounds to overcome this problem have been in use since 1940. DDT was used extensively to eradicate mosquitoes and in pest control in agriculture. Other chlorinated hydrocarbons, such as dieldrin, aldrin, endrin, lindane, and heptachlor followed, but the use of these substances to kill nematodes and other insect pests affected other soil fauna. Birds feeding upon affected insects were themselves poisoned and their reproductive capacity reduced (Westlake 1984). Some organo-chlorine compounds persist in the soil for up to ten years and many insects have developed an acquired resistance to them. Contamination at works where these chemicals were handled and spillages on farms must be expected to last as long.

Organo-phosphorous compounds such as parathion and malathion are used as systematic insecticides; these are sprayed on the plant or soil and taken up by the plant where they are ingested by the insect pest which is then killed. They are also very toxic to higher animals and vertebrates.

Synergists, such as piperonyl butoxide, are also insecticides which inhibit the nervous system of insects and at the same time knock out the metabolic functions of the mixed function oxidases of the invertebrate hepatopancreas. Unfortunately, they can also affect the function of the mammalian liver. Derivatives of carbamic acid are much less toxic to higher animals.

Herbicides are widely used in agriculture; many of them are specifically toxic to plants and are not a problem for animals. However, certain herbicides contain dioxins which are dangerous. The substances used as herbicides frequently include dinitro-orthocresol (DNOC) which inhibits the generation of adenosene-triphosphate in plant cells resulting in death of the plant tissue (Hance 1980).

Trialkyl-tins are used as fungicides and molluscicides in paints from which they are slowly released. Similar organo-tin compounds are used as polymerization catalysts and plasticizers in industry, but only the trialkyl-tins are stated to be hazardous.

Inhibition of photosynthesis by simazine is used to control plant pests. It is stated to be completely harmless to animals and is applied to the soil from which it is taken up by the plant roots. Paraquat attacks broad-leaf green plants; it is very poisonous if swallowed, but soil bacteria rapidly destroy it.

Auxins are compounds which imitate the natural growth-regulatory substances. They lead to distorted growth and death in dicotyledonous plants, whilst monocotyledonous plants are much less affected. In general, their persistence is not great and they have not been found to cause contamination. However, toxic by-products, dioxins, are produced in their manufacture and have caused much public concern about auxins used in agriculture and forestry. Their use has been discontinued in North America, but not as yet in the UK.

Any abandoned works which processed these compounds, or warehouse which stored them, should be carefully checked for their continued presence. Some of the residues may be strongly adsorbed, but leachates which contain them are very toxic to fish and leave a strong taste and unpleasant odour in water. DoE recommends the elimination of these substances by high-temperature incineration with appropriate control of effluent gases, rather than consigning them to landfill (Lindfors 1980). Dahl (1986) describes in detail 21 pesticide compounds present on a site in California. Dibromochloropropane (DBCP) was the most common substance identified.

Dutch experience (Table 4.1) suggests that less than 0.1 mg/kg of pesticides could be regarded as uncontaminated. If 2.0 mg/kg were present then further investigation is called for and the necessity for a cleansing operation begins above 20 mg/kg. Figures for individual chlorinated pesticides are slightly lower.

Phenols

Phenols and phenolic liquids are used in, and result from, many industrial processes. In large amounts phenol is a nerve poison. It can be readily absorbed through the skin where it produces local anaesthesia and caustic burns. Since 1974 phenols have been categorized as 'special wastes' in Britain and may only be disposed of on specialized sites.

Phenols may be dispersed in soils and underlying rocks, or be present in subsurface tanks and pipes not removed when the site was cleared. If they are encountered during restoration, specialist contractors should be called for the disposal operation. One of the major sources of phenols and similar aromatic compounds is from the carbonization of coal at a gas, coke, or tar works site. They may also be present in tarry wastes (DoE 1977*d*).

Oil refineries produce a wide range of petrochemicals which include phenols and phenolic compounds. Although most operations are enclosed, accidental spills and some official dumping may be expected. In the chemical industry, phenols may be expected as site contaminants where pesticides, plastics, and synthetic resins have been manufactured. Some phenolic residues arise also from the manufacture of plywood and from the cleansing of wool and hides.

Phenols (Table 4.2) should not exceed 5 mg/kg in soils of domestic gardens, allotments, or ground where pipes for drinking-water are to be laid. Other used can be permitted below 100 mg/kg, but above that level, a clean-up operation may be necessary. The Dutch (Table 4.1) recommend that below 0.2 mg/kg phenol in soil is normally a safe level, 1 mg/kg suggests further investigations are desirable, and above 10 mg/kg a clean-up operation is required. In water 15 μg/l phenol requires further investigation and where 50 μg/l is exceeded, the source of pollution should be found and removed.

Polychlorinated biphenyls

These substances have been used since the 1940s as dielectric fluids in transformers and capacitors, as hydraulic fluids, heat transfer fluids, lubricants, and vacuum pump fluids. Since 1973 they have been used only in transformers. Wastes arise from the manufacture and the demolition of electrical equipment containing polychlorinated biphenyls (PCBs). Although relatively inert, these substances are soluble in hydrocarbons. Those with few chlorine atoms can be broken down by bacteria, but as the number of chlorine atoms increases their degradability decreases (Sumner Report 1978), so they may be persistent; they are passed up food chains, where they present a health hazard to animals and man. Although DoE (1976*f*) recommends that they should be incinerated at high temperature with efficient scrubbers, some may have been spilt or consigned to landfill sites.

It is not known how limited dosages of PCBs would affect human beings but Leigh (1982) reported that a leaking transformer in a Japanese rice-oil factory poisoned 15 000 people. The symptoms included vomiting, swelling of the limbs, and cancer. The same article quoted a 'world expert' as saying that there is no dose which can be considered safe. The Dutch (Table 4.1) recommend that less than 0.05 mg/kg may be considered uncontaminated, but that any sample exceeding 1.0 mg/kg should be regarded as contaminated, and that concentrations in excess of 10 mg/kg require a cleansing operation.

Pulverized fuel ash

In modern power stations coal is pulverized and burnt suspended in a stream of air and the resulting ash comprises particles of fine sand (60 per cent) and silt. The ash has a specific gravity between 1.9 and 2.4, but about 5 per cent of it consists of very light hollow spherulites or cenospheres with an apparent specific gravity of 0.05. The pulverized fuel ash (pfa) is difficult to handle as it is easily blown about by the wind, and it will flow like a liquid if aerated. Consequently it is usually handled as a slurry in a 60:40 mixture with water. It is discharged into lagoons to settle from which it is ultimately removed and dumped. It has pozzolanic properties and tends to harden as it dries.

Compared with soil, pulverized fuel ash lacks a clay fraction and nitrogen, and is alkaline (pH 11–12) and contains soluble salts. Lagooning helps to reduce the salinity and the salts are fairly rapidly lost after dumping. The major problem with pfa is that it contains between 3 and 250 mg/kg water-soluble boron compared with the 2.5 mg/kg of boron in normal soils. It is generally accepted that 4 to 10 mg/kg B in soil is slightly toxic and more than 30 mg/kg is highly toxic, so the presence of boron presents problems in the use of pfa in restoration. However, the boron content is reduced by lagooning and by leaching once the ash has been dumped, after which the pfa can be a useful material for restoration, especially in areas like South Wales where the coal has a low boron content.

Pulverized fuel ash may be found on former power station sites and also on other derelict or former derelict areas (Fig. 4.2). At Peterborough pfa is being used to fill former clay pits. The pfa is brought by train from power stations at West Burton, Ratcliffe on Soar, and Drakelow 'C' in specially constructed wagons and unloaded by pressurizing them. As the ash is evacuated it is mixed with water and conveyed as a slurry to the former clay pits where it is allowed to settle. The water is recycled. When the pits are filled, compaction is reduced by tillage in the surface layer and topsoil washed from sugarbeet at the Peterborough works is used to cover the ash. Very satisfactory crops have been produced on this 'new' land (Fig. 4.3) (Central Electricity Generating Board n.d.; Barber 1975).

Quarry faces

A problem at all derelict quarries is the state of the quarry faces. Vertical or near-vertical faces are dangerous because people, particularly children, can fall down them, and even a single stone falling on

Fig. 4.2. Former gravel pit infilled with pfa, Trent Valley.

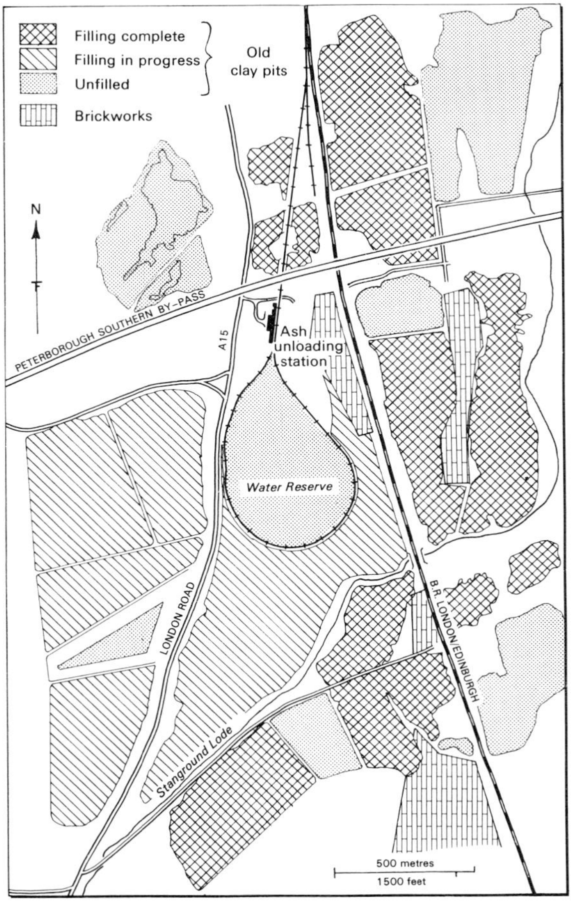

Fig. 4.3. Reclamation of former brickpits with pfa. (Reproduced by permission of the Central Electricity Generating Board, from *Land reclamation—Peterborough*.)

someone's head can cause a fatality. Larger falls of unstable stone may also occur. Regulations insist that there is a daily inspection of the faces for instability during the life of the quarry, but these inspections cease when quarrying ceases. Weathering continues and faces become weakened and unstable; eventually failure occurs and the face falls. Merely removing material and leaving a face unsupported induces stress relief; older quarrymen speak of the face 'talking' to them before falls occur. Considerable danger exists when bedding planes dip towards the quarry floor and it is even greater when clay is interbedded between rock beds as the clay may become lubricated with water and cause sliding.

The accumulation of water near the top of quarry faces must be prevented. It must be channelled away from the face or from any permeable strata and fissures where it could cause overloading of the face. Periglacial fractures of the upper parts of many quarries, especially chalk pits, make them particularly unsafe. Even where the quarry floor has been restored, the old quarry face itself may remain because there is insufficient space in which to lower the gradient between the edge of the quarry and a road. Consequently, in ironstone mining districts, and in the chalk-quarrying area of Essex, roads are frequently left high above the land on either side where quarrying has taken place. A safe angle for the Essex chalk would be 30° from the horizontal but insufficient material remains in the quarries to build a supporting embankment against the old face (Culbard and Essex 1983; Miller 1983).

Opencast (strip) mining in the Appalachian region of the USA has left high cliffs where coal has been worked down the dip on steep hillsides until it was no longer economic to extract more. This form of dereliction is difficult to restore satisfactorily (p. 18). The problems of collapse of sand and gravel quarry faces have been referred to in *Flooded workings*, this chapter.

Radioactive substances

The 1986 accident at Chernobyl power station in the Ukraine dramatically drew attention to the danger of contamination by radioactive substances. Before this catastrophe, contamination by radioactive substances, other than at sites where atomic weapons have been exploded, was not a major problem. Relatively minor spills and leakages had occurred, which caused justifiable public concern. When the first generation of atomic power stations is phased out, and should nuclear fuel processing plants and military nuclear installations (in shipyards dealing with nuclear submarines) become obsolete, there may be problems. Stringent precautions have been taken, so radiation levels should not be excessive even within the perimeters of their sites. However, the shielded core of a reactor will have to be made safe within thick layers of concrete for the foreseeable future. The only radioactive materials likely to occur on derelict land are relatively low-level residues from industrial uses.

Approximately 84 per cent of the radiation exposure received by the general population comes from natural sources such as from rocks or outer space through the atmosphere. Only 2 per cent of the radiation received is derived from the weapons-testing activity of the last 38 years. About 13.5 per cent comes from medical, dental, and other equipment, and according to the National Radiation Protection Board at Harwell the disposal of atomic wastes adds only 0.01 per cent.

In spite of stringent precautions some dangerous fission products do escape into the environment. Some of them emit α- and β-rays but these have a very limited range in air and do not constitute much threat unless taken directly into the body. Others emit γ-rays and neutrons which have a long range in air and are more dangerous and may pass through several centimetres of lead. This emission of radiation gradually falls with time, the rate of decrease is measured in terms of half-life: thus half of the level of radiation of ^{90}Sr is lost in 28 years, and in 56 years it has dropped to one-quarter, and so on. Other half-lives are ^{131}I = 8 days; ^{91}Y = 58 days; ^{144}Ce = 284 days; ^{3}H = 12.3 years; ^{239}Pu = 24 000 years; some are ephemeral with extremely short half-lives.

Unfortunately, certain nuclides are metabolized by the same processes as harmless elements. Thus radioactive strontium behaves like calcium and accumulates in the bones where the radiation it emits can affect production of blood cells in bone marrow and induce leukaemia. There was considerable concern expressed about the levels of strontium fall-out during weapon testing. The most serious accident in Britain so far occurred at Windscale (now Sellafield) where 7.40×10^{14} Bq of ^{131}I and 0.25×10^{14} Bq of caesium and strontium were released. As a result all milk produced within the surrounding 500 km^2 was collected and dumped for several weeks (Ottaway 1980).

The present unit of radioactivity is the becquerel (Bq) which is equivalent to one disintegration per sec. Formerly the unit was the curie (Ci) which was based upon the radiation given off by 1 g of radium. 1 Bq = 3.7×10^{10} Ci. Both units are used in the recent literature. The effects of radiation upon plants, animals, and human beings depend upon the radiation absorbed. The unit employed to measure absorption has been the rem which has recently been replaced by the sievert (Sv) (1 Sv = 100 rem).

The International Commission in Radiological Protection recommends that no one should be exposed to more than 50 mSv per year and for trainees under 18 years of age this figure is reduced to 15 mSv per year (Health and Safety Commission 1985; Mellanby 1980). The definition of a radioactive substance is any material having an activity concentration of more than 100 Bq/g^{-1}. If it has been shown that there are radioactive substances on a site, the provisions of the Ionizing Radiations Regulations 1985 and the accompanying code of practice (Health and Safety Commission 1985) should be followed to protect site investigators from the risks of inhalation, ingestion, and skin contact.

There is no level of contamination specified as acceptable because the aim is to achieve the lowest level consistent with working practices. If a spillage occurs or radioactive materials are discovered, where the instantaneous dose rate exceeds or is likely to exceed 7.5 μSv h^{-1}, the Radiochemical Inspectorate must be informed and a clean-up by specialist contractors should take place without delay. (Advice may be obtained from the Hazardous Materials Service at Harwell.) Before any radioactive material is moved the Inspectorate requires 28 days notice. Hazards are incidents above the critical quantities and concentrations given in Statutory Instrument No. 1333 (Health and Safety Executive 1985).

Red mud

Red mud is a waste product from the preliminary stages of aluminium production. Much of it has been dumped in the sea, but there may be lagoons of red mud near former aluminium works dealing with bauxite. Sometimes it was convenient to mix seawater with the red mud to make a slurry before pumping it into lagoons. In this case magnesium in the seawater reacts with the sodium hydroxide and releases sodium chloride which adds salinity to the difficulties of revegetating the surface. It has been suggested that the sandy fraction should be placed over the fines, but in practice it may be easier to cover them with other available materials.

Rock waste

All forms of mining and quarrying involve removal of some rock which contains no marketable minerals. In some cases rock waste can be used as aggregate but in many quarries it accumulates where it is produced.

Colliery shale has already been considered (*Colliery waste*, this chapter) as it has particular problems associated with the dereliction of former coal-mining districts. Other rock wastes include the rejected material from slate quarries, non-ferrous metal mines, and china clay workings.

Slate quarries in western Britain have produced over 500 million tonnes of waste rock (BSI 1985).

This rejected material makes a big visual impact upon the landscape of the Snowdonia National Park and is an industrial intrusion into an otherwise attractive mountainous area. It is unlikely to be used as aggregate as it is situated far from areas of demand and transport costs would be prohibitively high (Gutt *et al.* 1974).

Former non-ferrous metal mines where high-grade ores are extracted do not produce large amounts of rock waste. However, low-grade porphyry copper ore mining produces large quantities. A copper mine on Bougainville Island produces 85 000 tonnes of rock waste and 80 000 tonnes of tailings per day; 25 billion tonnes of waste rock have accumulated in the copper mining areas south of Tucson, Arizona, with a further million tonnes added daily (Bradshaw and Chadwick 1980). The extent of dereliction caused by mining and dumping of rock waste is great. Any restoration is made more difficult by the semi-desert or desert climate in which these areas occur.

In Cornwall and Devon about 35 square miles have been affected by kaolin mining and it is estimated that there are 280 million tonnes of waste in the two counties (Gutt *et al.* 1974). Abandoned kaolin mines are steep-sided, water-filled holes alongside which stand the conical waste tips. These wastes are lacking in plant nutrients so they remain unvegetated as the White Alps (Bradshaw *et al.* 1975).

Rubble

One of the most characteristic components of derelict land is rubble. Demolition of buildings on derelict sites produces broken masonry, bricks and cement, fragments of concrete, tile, and slate. The rubble may vary from fragments of bricks to large slabs of concrete. Unless they have been removed, rubble will normally contain fragments of wood and other combustible materials as well as the remains of galvanized iron sheeting, pipework, and other metal components.

Clean rubble has a considerable value as fill, or to cover less desirable material before laying a new concrete surface. Material from derelict housing, churches, and office buildings comes into this category.

Usually, the aim of demolition is to provide a cleared, level site which can be re-used. Clean rubble may be moved around a derelict site and mixed with other material to produce a level surface. If so, it may overlie impenetrable concrete flooring slabs or infill subsurface structures. However, rubble from industrial premises may be contaminated by toxic materials used in the manufacturing process. The most suspect will be rubble from old smelters, gasworks, iron and steel works, metal processing or fabrication works, and engineering and vehicle manufacturing works. Brickwork or concrete from mineral products and chemical works may also be contaminated. Thus, the fabric of base-metal smelters is usually contaminated with the elements which were smelted there. The ores of copper, lead, and zinc also contain arsenic, cadmium, nickel, and mercury as well as traces of manganese, chromium, and selenium. The rubble from flues and chimneys will be most highly contaminated, but virtually all the debris will be contaminated to some degree. There may be elevated levels of most metallic elements, particularly those used as alloys in former steel and tinplate works and, when they are dismantled, the toxic dust liberated will be spread around the site with the rubble.

The floors of engineering works will probably be soaked in oils and other solvents; there may be deep vats or tanks let into the floor of metallifereous works for cleaning and electroplating and these may have leaked and contaminated the foundations.

On a former gasworks site brickwork from the retort houses and purifier sheds may be contaminated with by-products of the carbonization process or substances used for purifying the gas. The most highly contaminated material may have been pushed into the former gasholder well and sealed in, but older demolitions may have spread contaminated brickwork all around the site (ICRCL 1979*a*).

The remains of former chemical and paint works may contain a wider range of contaminants. The rubble may be contaminated by solvents which have seeped into the brickwork, been accidentally spilt, or been thoughtlessly disposed of on site. At former paint works, some of the pigments used for colouring may be present including salts of copper, zinc, lead, mercury, arsenic, chromium, and tin. There will be a mercury problem at chlor-alkali plants.

Distillation processes at chemical works produce sludges and spent solvents and the petroleum and oil refining industries produce oily residues and tars which may have found their way into the fabric of the buildings, especially into concrete floors and foundations.

In addition, former chemical works or power stations must be assessed carefully for the presence of asbestos. Part of the insulation of the buildings or plant may have been discarded with innocuous materials in demolitions before 1969.

Soil erosion

As well as contaminating the site where they were produced, toxic elements released by an industrial process may also damage the surrounding areas. This is particularly obvious around smelters for copper, lead, zinc, nickel, and arsenic. Vegetation may have been damaged and killed by the sulphurous smoke and acidic rain to leave the soil open to sheet and gulley erosion.

In the Lower Swansea Valley, over 400 ha were damaged in this way and over one million tons of soil eroded; around a set of smelters in Tennessee, there are 7000 acres of bare ground and a further 17 000 acres of severely damaged land. Around a Montana smelter, fumes injured Lodgepole pines and alpine firs 19 and 22 miles away, respectively, and rendered pastures unfit through arsenical impurities for 15 miles around (Bridges 1969; Treshow 1976).

When such areas come to be reclaimed, there may be only contaminated, stony subsoil material to be landscaped. It may be necessary to bring in fresh top-soil to ensure satisfactory growth of new vegetation.

Solvents

Solvents, cleaners, and thinners are used in many industries and a significant proportion may have been disposed as waste. Waste Management Paper 14 (DoE 1977*e*) suggested that a few tens of thousands of tons per year may become waste. In many cases it is possible to reclaim solvents (33 million litres per annum) after use but even so 17 million litres are discharged into the sea and 1.7 million litres are disposed of in landfill. The solvents used in industry may be broadly grouped as in Table 4.4.

On a derelict site, solvents are most likely to be found in pipes and tanks which remained below ground after demolition. Any buried tank discovered during restoration should not be entered by workmen and any liquids found should be examined before spillage occurs. The ground below some works may be saturated with solvents disposed on site. Waste solvents may be adsorbed on to domestic wastes in landfill; in sites which have been examined it has been found that there has been a rapid attenuation in the concentration of solvents with depth in the unsaturated zone below the landfill. Clay layers are particularly important in containing solvents; calcareous materials have a similar effect.

Table 4.4. Solvents used in industry

Adhesives	Mixed solvents, methyl ethyl ketone, and toluene residues in containers
Aerosols	Reject batches in containers
Engineering	Paraffin, white spirit, paint thinner, cellulose thinner (also phenols and cresols for degreasing)
Leather industry	White spirit and paraffin for degreasing; surface-coating lacquers and thinners
Solvent manufacture	Still residues—little waste because of recycling (also pharmaceutical solvents)
Motor vehicle manufacture	Xylene, toluene, white spirit, kerosene containing dissolved oils, greases and waxes from protective coatings and dirt. Waste paints, thinners, and cleaning solvents
Nitrocellulose	Certain paints based on nitrocellulose—disposal requires at least 50-fold dilution and dispersal where heating is unlikely
Paint manufacture	Wastes from cleaning plant and rejected batches of paint, include resins, binders, fillers, and still residues. Sludges containing organic and inorganic pigments based on heavy metals, and biocides and pesticides based on mercury. Water-based paints containing wetting and antifoaming agents
Shipbuilding and repair	Paint and metallic residues from scraping or shot-blasting hulls—may contain copper, zinc, lead, chromium, tin, mercury, or arsenic. Up to 20 per cent of paint is antifouling which may contain cuprous oxide or alkyl tin compounds
Plastics	Wastes of residues in stills and reactor vessels which contain solvents
Printing	White spirit, kerosene, aliphatic hydrocarbons are used for cleaning. Ethyl acetate, isopropanol, ethanol, *n*-butyl- and isobutyl-acetate are used in photo-engraving and colour printing. Wastes include cadmium, chromium, nickel, lead, tin, zinc
Telecommunications	Aromatic and aliphatic alcohols and ketones are used for cleaning or as paint thinners

(This table is Crown Copyright and reproduced here by permission of the Department of the Environment, from DoE 1977*e*.)

The wide range of solvents used presents many problems when a contaminated site requires cleansing. Waste Management Paper 23 (DoE 1981) lists toxicity, flashpoint, corrosivity, and carcinogenicity levels.

Of the solvents mentioned, materials may be considered to be uncontaminated if they contain less then 0.05 mg/kg of toluene, 0.02 mg/kg of phenol, 0.05 mg/kg of xylene, and 0.01 mg/kg of aliphatic hydrocarbons. The surveyor should make further investigations if more than 3 mg/kg of toluene,1 mg/kg of phenol, or 5 mg/kg of xylene, and aliphatic hydrocarbons are present. A cleansing operation will probably be necessary for levels of 30 mg/kg toluene, xylene, and aliphatic hydrocarbons of 10 mg/kg of phenol. These trigger concentrations (Table 4.1) are derived from Moen *et al.* (1986) based on guidelines of the Ministry of Housing, Planning, and the Environment (Netherlands) (1983).

The widely used aliphatic hydrocarbons are not excessively toxic; aromatic and cyclic hydrocarbons are usually more dangerous, especially benzene (benzol) for which the maximum allowable concentration in the atmosphere is 25 mg/kg for an 8-h exposure. Alcohols, ketones, and esters are often irritants but are not usually highly toxic, but the maximum allowable concentration of carbon disulphide is 10 mg/kg for an 8-h day (Am. Conf. Govt. Indust. Hygienists 1974).

Spent oxides

The spent oxide used in the purification of town gas is a characteristic contaminant of former gasworks sites. After production in retorts, the gas was passed through an ammonia scrubber to remove hydrocyanic acid, carbon dioxide, and hydrogen sulphide. The remaining traces of hydrogen sulphide and cyanides were removed by passing the gas through trays holding ferric oxides. The sulphur from the hydrogen sulphide formed sulphide of iron and some sulphur. The strong acidity and the cyanides inhibit the growth of plants and may be hazardous for animals and human beings.

The provisional trigger concentrations (Table 4.2) for free cyanide, complex cyanides, and thiocyanate in soil materials is 50 mg/kg for all uses but a level of 500 mg/kg is permissible around factories where no landscaping is intended. Sulphur may reach 1000 mg/kg before it reaches the trigger concentration suggested by ICRCL (1983). In the Dutch guidelines (Table 4.1) further investigations are necessary when the free cyanide concentration exceeds 10 mg/kg and for complex cyanides 50 mg/kg. Free cyanide exceeding 100 mg/kg and complex cyanides exceeding 500 mg/kg indicate that a cleansing procedure is necessary.

Subsidence

The old pillar-and-stall method of mining did not lead to excessive subsidence as sufficient coal, iron ore, or salt was left behind to support the roof. However, the longwall method extracts nearly all the mineral seam and the roof is allowed to fall in. This has led to widespread subsidence in areas where coal has been mined.

Although subsidence has been minimized under important buildings and major built-up areas, there has been substantial subsidence of the intervening rural land. Many houses have been badly cracked and have had to be demolished; roads, railways, and canals have had to be reconstructed to maintain grades or levels. Disruption of drainage is probably the most obvious sign of subsidence in country districts; the appearance of flooded areas (flashes) and the malfunctioning of field drains all point to subsidence.

When salt was mined in Cheshire, sufficient was left undergound to support the roof, but the galleries of the old mines were subsequently flooded; when brine was pumped many of the pillars were dissolved and subsidence was common. Some of the surface hollows formed by subsidence in Cheshire have been used for dumping chemical wastes, particularly from alkali works (Wallwork 1974).

Sulphates

Natural calcium, magnesium, and sodium sulphates occur in the Keuper Marl, Oxford and Kimmeridge Clays, Lower Lias, and London Clay. Also sulphate-rich ground waters occur sufficiently near the soil surface in many semi-arid and arid countries such as Tunisia and Bahrain, for gypsum to be deposited in the soil as the moisture evaporates. Sulphates are also formed by the oxidization of pyrite in coal-shale waste, and sulphates may be released from oil shale residues, pfa, and any builders' rubble which contains an appreciable amount of plaster. Many industrial processes are sources of sulphur and sulphates on derelict and contaminated sites: a former gasworks site will commonly contain a high content of sulphides and elemental sulphur; blast furnace wastes can contain sulphur or sulphate; waste from chromite smelters contains crystalline sodium sulphate and landfill sites commonly contain putrescent materials which contribute sulphur compounds which may be corrosive to concrete structures.

Sulphate in soil may attack concrete by crystal growth and the production of a sulpho-alumino-silicate mineral, ettringite, which causes a substantial increase in the volume of the concrete which leads to disruption. Rusting of the iron reinforcing rods follows and the whole structure is greatly weakened. The severity of the problem depends on the supply of sulphates and a source of moisture to carry it into the concrete structures. If a high ground-water level brings the capillary fringe near to, or above, the level of the soil, solutions can migrate upwards into the structure and when the water evaporates it leaves the salts behind. Two cases of sulphate contamination in Lancashire are given by Barnes (1986).

The concentration of sulphate in ground waters is measured as grammes per litre of sulphur trioxide, but this gives an indication only of their aggressiveness as solubilities vary considerably. Calcium sulphate has solubility of only 1.2 g l^{-1} (SO_3), but magnesium and sodium sulphate have 150 to 200 times this value. With soil samples the total amount of sulphate can be determined by extraction with hot dilute hydrochloric acid. The calcium sulphates can be distinguished from the potentially more dangerous magnesium or sodium sulphates by extraction by a 2:1 water to soil ratio. Expressed in terms of g l^{-1} concentration of SO_3 sites may be classified into five categories with differing requirements for the cement to be used in concrete on the site.

A site is classified on the concentration of SO_3 (g l^{-1}) in the 2:1 and 1:1 extract:

Site classification	Concentration of SO_3 (g l^{-1})	
	In 2:1 extract	In 1:1 extract
1	0–1.0	0– 1.0
2	1.0–1.9	1.0– 2.5
3	1.9–3.1	2.5– 5.0
4	3.2–5.6	5.0–10.0
5	> 5.6	> 10.0

The requirements for concrete exposed to sulphate attack in the five site categories are given in Table 4.5.

According to ICRCL (1983) (Table 4.2) sulphur may be tolerated in soils up to concentrations of 1000 mg/kg for all uses, but above that level further investigations should take place. The trigger concentration for sulphides is set at 250 mg/kg and for sulphates at 1000 mg/kg, below which levels land may be considered free from contamination. A threshold of 2000 mg/kg can be allowed for soils in open spaces or landscaped areas.

Zinc

Zinc was used in brass from pre-Christian times and in Europe as a metal after about 1700. A modified brass with increased zinc was developed by Muntz in 1832 and used for sheathing wooden ships. In 1837 the process of coating sheets of iron with zinc to produce galvanized iron caused the demand and world production of zinc to increase by a factor of 70 between 1837 and 1890. Zinc is used in other alloys also, such as those used in die-castings; zinc chloride is used in soldering and to prevent decay in wood; zinc sulphate has been employed in dyeing and glue manufacture. Plates of metallic zinc are used in the printing industry and zinc is used in the pharmaceutical industry both as a constituent and as a container (toothpaste or ointment tubes), as well as for dry battery cells.

Zinc is obtained from its sulphide ore, sphalerite (ZnS), which occurs in similar situations to copper and lead with which it is frequently associated. It is mined from deep mines, like lead, and no low-grade deposits have been found similar to the porphyry copper ores. Canada, USSR, USA, and Australia are the leading producers of zinc at present, though many lead–zinc mines once operated in Britain, particularly in Wales, and Derbyshire in England. At the mines, the associated dereliction and contamination problems are similar to those discussed under *Lead*, this chapter. The zinc ore is crushed and concentrated, as for lead, to give a high-grade product for the furnaces with about 80 per cent zinc. Contamination around the smelters affects soils, vegetation, and crops. Jennett and Wixson (1972) reported that the soil content of zinc (and lead) decreased away from mines and smelters in south-east Missouri until background levels were reached 5 to 6 miles away. To limit the spread of contaminating dusts, they recommended that ore should be sprayed with water to keep it moist, vehicles carrying ore should be covered, and filled tailings ponds should be covered with uncontaminated soil to seal in any contaminants.

The slag heaps of waste material produced by eighteenth- and nineteenth-century smelters tend to be dark grey in colour with a fairly uniform sandy texture. Zinc waste tips are almost devoid of plant growth. Indeed, some older tips in the Swansea Valley

Table 4.5. Requirements for concrete exposed to sulphate attack (Building Research Establishment (BRE) 1981). (Recommendations are for concrete in a near-neutral ground water—for acid conditions refer to BRE Current Paper CP 23/77)

Class	Concentrations of sulphates expressed as SO_3: In soil: Total SO_3 (per cent)	In soil: SO_3 in 2:1 water:soil extract (g/l)	In ground water (g/l)	Type of cement		Requirements for dense fully compacted concrete made with aggregates meeting the requirements of BS 882 or 1047: Minimum cement content[1] (kg/m^3)	Maximum free-water/cement ratio[1]
1	Less than 0.2	Less than 1.0	Less than 0.3	Ordinary Portland cement (OPC) Rapid Hardening Portland cement (RHPC) —or combinations of either cement with slag[3] or pfa[4] Portland Blastfurnace cement (PBFC)	Plain concrete[2]	250	0.70
					Reinforced concrete	300	0.60
2	0.2 to 0.5	1.0 to 1.9	0.3 to 1.2	OPC or RHPC or combinations of either cement with slag or pfa PBFC		330	0.50
				OPC or RHPC combined with minimum 70% or maximum 90% slag[5] OPC or RHPC combined with minimum 25% or maximum 40% pfa[6]		310	0.55
				Sulphate Resisting Portland cement (SRPC)		290	0.55
3	0.5 to 1.0	1.9 to 3.1	1.2 to 2.5	OPC or RHPC combined with minimum 70% or maximum 90% slag OPC or RHPC combined with minimum 25% or maximum 40% pfa		380	0.45
				SRPC		330	0.50
4	1.0 to 2.0	3.1 to 5.6	2.5 to 5.0	SRPC		370	0.45
5	Over 2	Over 5.6	Over 5.0	SRPC + protective coating[7]		370	0.45

[1] Inclusive of content of pfa or slag. These cement contents relate to 20 mm nominal maximum size aggregate. In order to maintain the cement content of the mortar fraction at similar values, the minimum cement contents given should be increased by 50 kg/m^3 for 10 mm nominal maximum size aggregate and may be decreased by 40 kg/m^3 for 40 mm nominal maximum size aggregate.

[2] When using strip foundations and trench fill for low-rise buildings in Class 1 sulphate conditions further relaxation in the cement content and water/cement ratio is permissible.

[3] Ground granulated blastfurnace slag. A new BS is in preparation.

[4] Selected or classified pulverized fuel ash to BS 3892. A new BS superseding BS 3892 is in preparation.

[5] Per cent by weight of slag/cement mixture.

[6] Per cent by weight of pfa/cement mixture.

[7] See BS CP 102:1973: Protection of buildings against water from the ground.

contained up to 11 per cent zinc and it was suggested that they should be transported to Avonmouth to be resmelted by modern methods (Fig. 4.4).

Typical levels of zinc in the environment, other than in metalliferous zones and contaminated areas are: igneous rocks 70 mg/kg; shales 95 mg/kg; sandstones 16 mg/kg; and limestones 20 mg/kg. Soils may be expected to contain up to 50 mg/kg when uncontaminated (Skinner 1969). Between 2700 and 9000 mg/kg were found in eroded soils immediately

Fig. 4.4. Zinc wastes in the Lower Swansea Valley, 1974.

downwind of former zinc smelting operations in the Lower Swansea Valley (Hilton 1967).

Zinc will probably be present around nearly all metal works, especially those processing copper, lead, tin, arsenic, titanium, and mercury. It will be a contaminant of scrap-yards and a minor contaminant of coal carbonization plants. It occurs in the effluent from tanneries and can be a major problem in the use of sewage sludge on agricultural and horticultural land. Soils spread with sewage sludge on the Beaumont Leys estate in Leicester, contain total zinc levels between 69 and 3050 mg/kg and total zinc was between 42 and 1070 mg/kg on areas irrigated by effluent. The levels of available zinc were lower, between 3 and 1980 and between 3 and 635 mg/kg, respectively (Fig. 3.16) (Heeps and Pike 1980; Wood and Ross 1980).

Zinc is used to make printing plates; in processing, plates are etched and cleaned with acids and solvents which may have been dumped locally on some sites. Since 1974 such wastes should have been consigned to landfill where they would be diluted and adsorbed on other wastes. Zinc similarly is a component of wastes from pharmaceutical works and paint works.

In Britain, the trigger concentration (Table 4.2) for plant-available zinc is 130 mg/kg; the Dutch (Table 4.1) accept 200 mg/kg. Further investigations should take place at 500 mg/kg and clean-up measures over 3000 mg/kg (Table 4.1). Zinc is phytotoxic but not a major threat to animals and human beings. The JURUE* survey of Walsall indicates that it has reached danger levels in some areas where metal-working occurs or formerly took place (Haines 1981). The Dutch guidelines (Table 4.1) require further investigation if zinc concentrations in surface or ground water exceed 200 μg/l; over 800 μg/l a cleansing operation is thought necessary. In the USA a TLV/40 of 120 $\mu g/m^3$ is the maximum acceptable level of zinc in workroom air (Am. Conf. Govt. Indust. Hygienists 1974).

*(Joint Unit for Research on the Urban Environment (University of Aston, Birmingham).)

5 Survey and assessment

Assessment can begin once a decision has been made to reclaim an area of derelict or contaminated land (Fig. 5.1). The stages described in this chapter take place within the legal and financial framework outlined in Chapter 7 for England and Wales; these may be slightly different elsewhere. Since the proposed land use may influence decisions about restoration, the surveyor will need to know about it. In general terms, the surveyor of derelict land must:

1. Define the site accurately;
2. Ascertain the uses envisaged;
3. (i) attempt to ascertain what is on a site,
   (ii) what are the problems, and
   (iii) what must be done to solve these problems, before the site can safely be used again.

In view of the range of problems likely to be encountered assessment may require the collaboration of experts in several disciplines. To avoid overlap or omissions they must follow a logical method of approach. There is no standard code of practice at present for the investigation of contaminated land, but several schemes have been suggested (Kelly 1980; Dawson and Brown 1981; Wilson and Stevens 1981; Bell *et al.* 1983; Kooper and Mangnus 1986). Assessment of the engineering suitability of the ground is covered by BS 5930 (British Standards Institute 1981); another volume will deal with contaminated land (Lord 1983). The approach described here parallels the code of practice outlined in the discussion document EPC/47 (BSI 1983). It aims to provide a framework that will assist the surveyor of derelict and contaminated land to provide an accurate and realistic appraisal of the problems, and the basis for a strategy of reclamation.

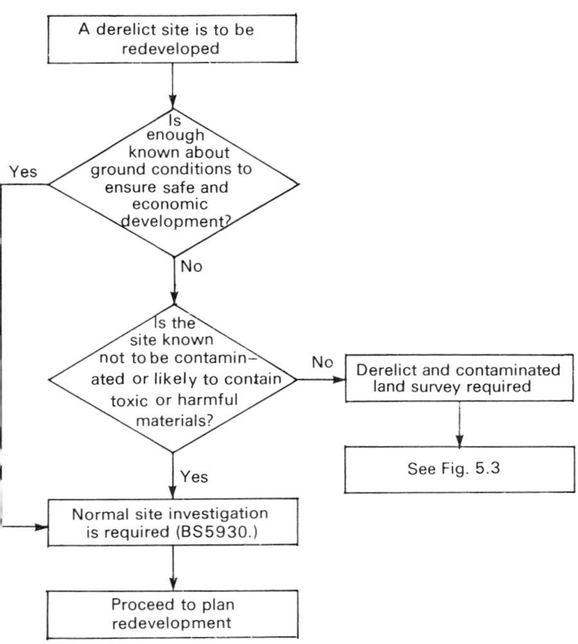

Fig. 5.1. Flow chart for the initial decision.

There are three main stages in assessing derelict and contaminated land and planning restoration (Stages 1–3 in Fig. 5.2).

1. **Preliminary assessment** of the problems of the site by a document search, site visits, and the laboratory examination of a small number of samples.
2. **Planning the sampling and measurement.** The strategy for further investigations is based on the preliminary assessment. At this stage it may be necessary to bring in expertise from several disciplines. Not all specialists are needed for every site, but the list includes hydrogeology, pedology, chemistry, biology, physics, topographic survey, and quantity survey. Individually or in co-operation, they plan a sampling programme to ascertain the distribution of materials and pollutants horizontally and vertically throughout the site.
3. **Specialist surveys** take place according to the plan and a comparison is made with the threshold values in the available guidelines for the use of sites after reclamation. A restoration programme is then drawn up to meet the client's requirements and environmental safety standards.

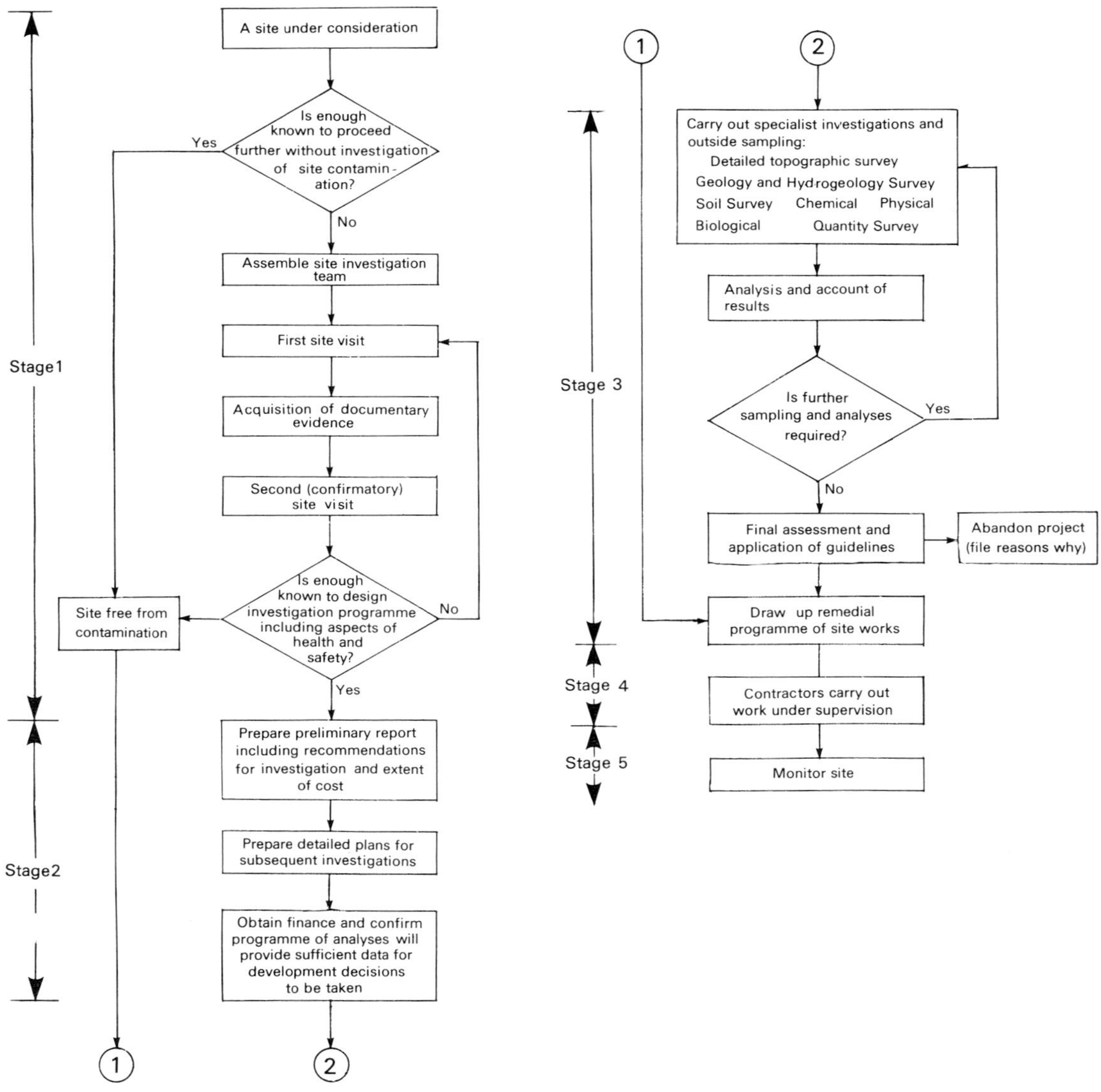

Fig. 5.2. Flow chart for the investigation and assessment of derelict land and contaminated land.

4. The **engineering works** necessary to achieve the restoration.
5. Finally it may be necessary to *monitor* the site for a period of time after restoration.

Although Fig. 5.2 sets out the procedure in a continuous flow diagram, it may be thought of as a series of loops. Each step depends on the adequacy of the preceding work. If this work was inadequate, it will be necessary to return for additional or better information. If the preliminary assessment indicates there are no problems of contamination, it will be possible to draw up a remedial programme without the expense of a full-scale sampling programme and the subsequent detailed laboratory analyses.

Preliminary site assessment

First site visit

If it is known that an industrial site is to be cleared, a site visit should take place before demolition. The visit will be more informative if the surveyor is accompanied by a person familiar with the layout of the works and the processes carried out. Even if the works have already been demolished the first site visit will be helpful in giving the surveyor a clear impression of the site upon which to base his findings. Before the first site visit the surveyor should collect a detailed site plan, district maps, air photographs, and any available maps of geology and soils. Where necessary, permission to enter should be obtained from owner and/or occupier. The site should be walked over and the boundaries of the site verified. Existing buildings should be compared with the detailed plan, and differences or omissions marked. The alignment of roadways and services such as sewers, gas, telephone, water, and overhead electricity transmission lines should be checked. The nature of the adjacent property should be noted and any likelihood of it being affected by the proposed works. The access to the site should be investigated, noting especially limitations for heavy demolition or construction traffic on the existing roads, bridges, and services.

The surface configuration of the site should be noted carefully and compared with the maps and available air photographs: the need for excavation and/or fill should be noted. The composition of any wastes deposited on the site should be assessed qualitatively and their distribution marked on a site plan. Direct evidence of contamination should be noted, and the presence and level of standing water in wells or ditches with any other surface-water bodies, springs, and seepages. The nature and vitality of vegetation growing on wastes or natural soil should be recorded.

It is very useful to photograph features of the site before, during, and after demolition. Such photographs assist off-site discussions: if buried structures affect subsequent restoration work, the photographs may indicate the dimensions and location of the problem.

It may be helpful to collect some samples during the initial site survey. Samples from the surface or from shallow pits may confirm the nature of materials on site. A few preliminary samples should be taken of any suspicious materials, liquids, and gases on site and analysed, though an obviously contaminated site will require a more thorough sampling in the next stage.

Acquisition of documentary evidence

Documentary evidence may be obtained from both official and unofficial sources. When a works is about to close for site redevelopment, the occupier or owner will usually provide information about it. This may include deeds, specifications, building development plans, business records, and old photographs, all of which may assist site assessment. If the site has been abandoned, it will be necessary to approach former owners or workers. The latter may reveal practices, unknown to the management, which may be of importance when it comes to assessing contamination.

Where it is no longer possible to trace owners or workers, relevant papers, such as sale catalogues, newspaper articles, and other unpublished material, may be housed in the local (archive) record office, museums, or university libraries. Technical journals and studies in industrial archaeology may describe buildings and other structures similar to those on a derelict site. Business directories, especially those published at regular intervals in the past, can be very helpful in establishing the succession of activities on a site.

Maps and plans of the area may be inspected at local government and record offices as well as some libraries. Historical sources before 1836 include enclosure maps and tithe maps. Between 1843 and 1894 nearly 400 town plans were surveyed at a scale of 1:500 and from 1853 onwards there are the plans of the Ordnance Survey. The successive revisions of the Ordnance Survey maps and plans might indicate the growth or contraction of works.

Systematic air photography began during the Second World War and photographs of many areas of Britain are available from the mid-1940s onwards. In some cases air photographic cover is available for several dates and may reveal changes in the use of a particular site. Air photography can be used simply as a picture of the ground surface, but air photographic interpretation may reveal considerably more information about a site, and photogrammetric techniques can be used to produce an accurate map and representation of relief.

Documentary evidence of industrial activity may be obtained from the National Register of Archives, the Business Archives Council, the Panel for History of Industry, and the individual industries themselves, such as the Archives Panel of the Institute of Civil Engineers. The Factory and Alkali Inspectorate has information on the processes used at particular premises, and the Regional Water Authorities have data on pollution of streams and ground water in their catchment areas. Advice on the underlying geology may be obtained from the Institute of

Geological Sciences, including the logs of borings and wells. The Mining Record Office, the Inspectorate of Mines and Quarries, and the National Coal Board have information on sites where mineral extraction has taken place. A list of such preliminary sources of information has been compiled by the Transport and Road Research Laboratory, Report LR403 (Dumbleton and West 1976) and aspects of the historical investigation of site use are discussed by Dumbleton (1980).

Some local authorities with extensive derelict land have instituted 'derelict land units' staffed by experts, for example, Lancashire where the County Council and the Greater Manchester Council jointly support a reclamation team; the Welsh Development Agency supports one in South Wales. There is also a Hazardous Materials Service operated by UKAEA, Harwell which is equipped to deal with emergencies and provide routine advice and assistance on a wide range of pollution problems as well as radiation. There is also an Interdepartmental Committee on the Redevelopment of Contaminated Land (ICRCL) in the Department of the Environment to assist local authorities with reclamation problems and to act as a clearing house for information about derelict and contaminated land.

Confirmatory site visits

The first site visit gives the surveyor an impression of the site geography and the obvious problems. When he has completed his documentary search and interviewed management and former workers the surveyor should visit the site again.

By this time he should have all the readily available information, and in the light of his experience, he should be able to identify which parts of the site are contaminated. He should now look for mineshafts or tunnels or smaller underground structures such as storage tanks, sewers, and culverts which may be evidenced by manholes. They may be older than the buildings on the site and their structural condition may be poor. If necessary, their replacement must precede other restorations.

The dimensions of heaps, holes, and lagoons should be recorded and the surveyor should look for ground disturbed by subsidence or settlement. He should estimate the quantity of material in waste heaps and their potentials or hazards and look carefully for the presence of asbestos or any other chemical wastes which could be a potential danger to site workers during restoration.

It may be possible to complete the preliminary investigations with one confirmatory visit or the surveyor may have to return several times to check specific points.

The preliminary investigation report

The conclusions of the preliminary work are presented in a **preliminary investigation report**. This should be clear and positive in its recommendations and should present and analyse all the information obtained from site visits, with documentary evidence revealed by maps and literature and any verbal reports. It should present evidence for contaminants and other adverse factors with recommendations for further investigations where necessary. The report should be illustrated with plans and photographs.

This report forms the basis for any further work required upon the site. It may come to one of four main conclusions:

1. The site is free from problems and is satisfactory for the proposed use.
2. The site has problems; the report presents estimates of the cost of correcting these so that the site can be used for the proposed development.
3. Further investigations should be made before a decision can be made about the suitability of the site.
4. On the evidence available, the site is unsuitable for the proposed use.

If it is certain that a site is free from contamination, then restoration may proceed. If contaminants are present and they can be made safe for the proposed development, then the preliminary report should recommend a **detailed site survey**. The detailed survey should provide information about any features of the site which require accurate quantification and particularly on the presence of toxic or potentially dangerous materials which would present hazards on site workers. The preliminary report will suggest the number, depth, and position of sampling points, and recommend the tests appropriate to the range of contaminants suspected.

Design of subsequent investigations

The preliminary investigation will have indicated the nature of the site and its most likely contaminants. If it recommended a detailed survey, then the sampling strategy must be planned to accommodate the known problems and identify the unknown ones. First, if any 'special wastes' are revealed by the preliminary survey, these must be dealt with or removed to a place where they can be accommodated safely. They might include asbestos lagging, toxic sludges, and liquids from lagoons or tanks. If possible, toxic materials should be made harmless on site to avoid their spread through accident during transit.

On a large site the investigating team will require a portable site office or caravan linked to existing services. This will provide a clean, dry working space for site plans to be laid out, where amendments can be made, or revised sampling positions marked. Equipment and samples may be stored and shelter is available for staff during inclement weather. Derelict sites are potentially dangerous areas: where contaminants are known to be present, protective clothing should be worn during site investigations, first aid equipment should be available and other precautions taken as required by the Health and Safety at Work Act, 1974.

The sampling strategy depends upon the purpose of the investigation and the proposed intensity of land use after reclamation. A number of sampling strategies may be used (Bell *et al.* 1983). Samples may be collected on a single visit, analysed for obvious contaminants, and stored for future analysis if required. Alternatively, sampling may be achieved in stages; the results of the first stage are used to plan a subsequent phase of increased sampling intensity. Certain contaminants, such as leachates or gases, can only be assessed by sampling at regular hourly, daily, weekly, or monthly intervals as appropriate.

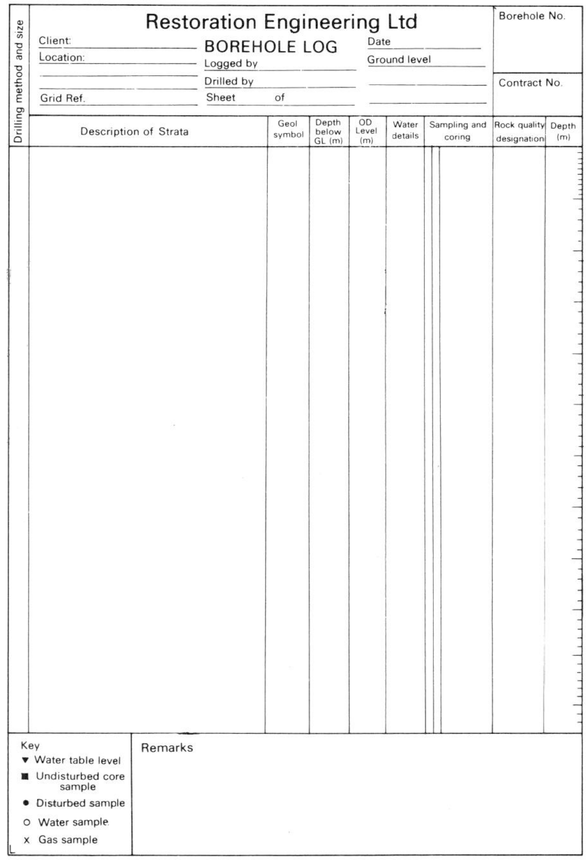
Restoration Engineering Ltd
BOREHOLE LOG
Drilling method and size
Client:
Location:
Grid Ref.
Logged by
Drilled by
Sheet of
Date
Ground level
Borehole No.
Contract No.

Description of Strata	Geol symbol	Depth below GL (m)	OD Level (m)	Water details	Sampling and coring	Rock quality designation	Depth (m)

Key
▼ Water table level
■ Undisturbed core sample
● Disturbed sample
o Water sample
x Gas sample
Remarks

Fig. 5.3. Sample form for recording data on derelict and/or contaminated land.

Sampling soils and waste material, liquids, and gases

The sampling sites for soils and waste materials should be located carefully according to the sampling plan and the positions marked with clearly visible stakes, each of which has been allotted a code number. The layout of the sampling plan should not take place too long before sampling, otherwise markers are lost or vandalized. Inevitably there will be problems, for example, impenetrable foundations may necessitate the relocation of a sampling point. Any changes should be marked on the plan. Data are normally recorded upon *pro forma* sheets which helps to make the collection of information uniform (Fig. 5.3). Description of the material down to undisturbed rock strata is desirable with sketch sections upon which the depth sampling positions are marked. These descriptions and sketches are to assist the subsequent interpretation of analytical figures. The containers for samples should be clearly marked with the location and depth at which the sample was taken.

To avoid contamination of the samples, they should be taken using clean, stainless steel tools, which should be cleaned between each sample. Surface samples may be taken most easily with hand tools, and on a small site where little contamination is suspected 1-m pits may be excavated by pick and shovel. Where the number of pits becomes greater than 10 to 12, or deeper pits are required, the use of a tractor-mounted back hoe becomes economical as deeper digging is facilitated.

As material is brought up by the back hoe, large lumps should be broken open and sampled with a stainless steel trowel as described previously for surface samples keeping a careful check on the exact depth and location of the sample. It is also possible to use the back hoe to break unreinforced floor slabs. Greater depths may be reached by digging a ramp down into the ground being sampled.

Deeper sampling can be achieved by larger machinery, usually a track-mounted excavator which may be able to break reinforced concrete slabs. Where deeper investigations are carried out it may be necessary to install shuttering to prevent collapse of the sides of the excavation and personnel should always be on their guard for gas emissions in excavations.

Sampling at depths greater than 5 m necessitates the use of boring equipment. Although the need for large excavations is avoided, the exact depth of samples from a borehole is difficult to ascertain and some methods involve a high risk of mixing materials from different horizons. It is possible to take core samples from consolidated materials using a rotary drill which introduces a casing to enclose a core of the material. Note that a deep borehole may puncture an impermeable floor below wastes and allow toxic fluids to contaminate aquifers. Excavations and borings should be carefully back-filled after sampling to prevent the migration of fluids or gases, or access of air into materials which might prove combustible.

Equipment for site investigation is fully described in the British Standard Code of Practice for Site Investigations (British Standards Institute 1981).

Sampling patterns The following sampling patterns may be used to enable the surveyor to obtain representative samples from which reliable information can be gained upon which to base recommendations for restoration.

Simple patterns Sampling the surface material upon a site during the preliminary assessment may be accomplished with a hand tools from the top 20 cm along an X- or W-shaped sampling pattern. Samples should be taken at regular intervals, giving a picture of surface conditions over the whole area (Fig 5.4(a), (b)).

Systematic sampling A grid pattern is superimposed on the area and samples are taken from the points of intersection of the grid which may be square, rectangular, or diamond-shaped (Fig. 5.4(d)–(f)). Grid sampling has the advantage that an even cover of the site is obtained and it is possible to return to sampling positions should unforeseen problems occur. (An unlikely disadvantage of regular sampling could occur if the periodicity of sampling coincided with the variations in the ground, hitting all the highest or lowest points of contamination.) A grid sampling pattern is favoured by most surveyors as the whole area is sampled uniformly, the likelihood of

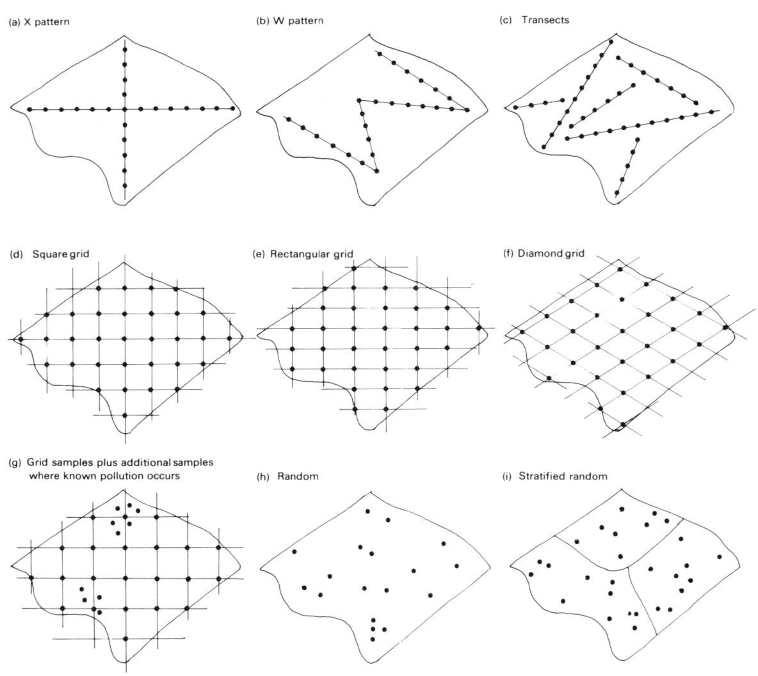

Fig. 5.4. Sampling designs.

finding previously unknown problems is good, and contaminated areas may be delimited (Bell *et al.* 1983).

Waterhouse (1980) suggested that a 95 per cent confidence level for the level of major contaminants is acceptable in practice. This means that the distance between samples may be extended on larger sites (Table 5.1). However, if intervals are more than 100 m, it may be difficult to interpolate with confidence between sampling points. Ideally, samples should be taken at the intersections of a 25 m grid, regardless of area, and it is suggested that the minimum number of samples from any one site should be between 10 and 20 (Smith and Ellis 1986).

Table 5.1. Size of site, grid, and number of samples (Waterhouse 1980)

Site area (ha)	Grid size (m)	Number of sample points
0.5	10	66
5.0	20	140
16.0	30	200

Extra samples should always be taken in areas where known pollutants occur, even though these points do not fall on the predetermined sampling grid (Fig. 5.4(g)).

Random sampling After placing a grid over the area concerned, random numbers are used to provide a series of randomly chosen sites for sampling. Although this sounds satisfactory, it frequently results in large areas of the site not being sampled and contaminants can easily be missed (Fig. 5.4(h)). However, many derelict sites can be subdivided into areas of similar characteristics, often of approximately equal size. Where this can be done, the surveyor could consider the use of a stratified random sampling plan (Fig. 5.4(i)). The advantages of this approach are that means, standard deviation, and error estimates can be calculated, but the random sample points can be difficult to resample should this prove necessary. In this method it is possible to vary the number of samples so that the most contaminated areas have the greatest sample density.

Transects Where the terrain is such that a grid sample plan is difficult to put into practice transects across known areas of contamination can be effective. A number of transects can be made into a flexible grid, the orientation of which can be varied using the experience of the surveyor. Samples are taken at fixed intervals along the lines of the transects (Fig. 5.4(c)).

Samples in depth As contamination of a site is a three-dimensional problem a sampling programme should include provision for collecting samples at various depths through the materials on the site. Surface samples may be taken with a trowel and pits may be dug to 1 m, but it is advisable to take some samples at 2 and 3 m depth. Where highly mobile contaminants are known to have affected a site, samples should be obtained from a borehole.

It should be assumed that all liquids found on a derelict site are toxic unless proven otherwise. Where the preliminary investigations have shown a strong likelihood of contaminants, protective clothing should be worn. Clean, clearly labelled glass or polypropylene bottles are normally used for liquid samples; the containers should be rinsed with the liquid before filling. Samples should be sucked-up by vacuum pipettes; mouth pipettes should never be used for sucking samples from the site into containers.

Samples should be transported as quickly as possible to the laboratory as they may deteriorate after collection. Certain samples may produce gases which could cause sample containers to burst. If this is suspected, vessels with pressure-relief valves should be used.

Gas samples can be withdrawn from a borehole or other confined spaces by the use of hand pumps. Glass or metal containers are used for samples as plastics and rubber are relatively permeable to gases and may absorb vapours. Unlike liquid samples, gas samples do not deteriorate rapidly, so the time taken to get them to the laboratory is not critical. The general rate of emanation of gases can be determined by placing a cowl over the ground; after equilibrium has been reached, measurement can be made of the accumulated gases. The presence of any gases and level of potential hazards can be determined with handheld equipment to an accuracy sufficient for safety control.

Aerial parts of plants may be sampled with secateurs or their roots may be dug out and washed free of soil and samples placed in labelled plastic bags. Microbiological materials require careful sampling with sterilized equipment. Many human pathogens are anaerobic bacteria which can survive in poorly drained soils. Land which has sewage sludge or effluent disposed upon it should be examined for bacteria responsible for tetanus, gangrene, food poisoning, and polio viruses. Sites where it is known that veterinary and pathological samples have been disposed of should be treated with considerable caution. Anthrax is known to persist for several decades. Staff should be meticulous about hygiene when collecting biological samples and the wearing of protective clothing should be based on the assessment of risk in any specific case.

Other problems which might come to light during the survey and necessitate special care are the presence of asbestos, heavy metals, phenolic liquids, tarry materials, explosives or radioactive materials, all of which require expert handling which should be engaged immediately.

Specialist surveys

Specialist surveys will normally be undertaken at appropriately detailed scales, generally greater than 1:5000. Specialists most commonly involved in surveying derelict land include land surveyors, photogrammetists, geologists, pedologists, chemists, physicists, ecologists, and quantity surveyors. A review of current site investigation methods was given by Hatayama (1986).

Topographic survey

On a complex site, a photogrammetrical map will be the most expeditious method of gaining information about relief and the distribution of materials. However, on smaller and less complex sites the land surveyor may measure the site and its features on the ground. In many cases, waste heaps conform to relatively simple geometric shapes, such as cones or embankments, with volumes that are easily calculated. This simplicity can be upset if contractors have extracted hardcore, leaving a jumbled landform of small hillocks and hollows. Features of former buildings above ground are fairly straightforward to measure but below ground the size of foundations and other structures can be difficult to estimate. Consequently, most contracts have clauses which take this into account.

Using stereopairs of air photographs it is possible to identify, classify, and accurately locate individual types of industrial dereliction. Although field checks are still necessary, the accuracy and speed of compilation is faster than working on the ground, mapping each material in turn. Several schemes have been used in Britain. One of the most comprehensive was developed in 200 km^2 in Yorkshire (see pp. 10–12) and was presented by Bush (1969) and Collins and Bush (1969*a, b*).

Air photographs can be used to construct an accurate relief model of a derelict site in more detail than can be shown on a conventional topographic map. If earth-moving is contemplated, the quantities of tipped materials on site and the amounts needed to be moved to achieve the final landform design may be calculated.

Remote sensors include sophisticated equipment such as multispectral scanners mounted in aircraft or satellites. Objects 80 m across are visible from satellites, but at lower elevations airborne scanners can sense objects 2.5 m across. Sensors in the near and middle infra-red have been used to map bare rock surfaces, soils, and materials on the surface of derelict sites and the use of ultraviolet is currently being investigated. Several hazardous waste materials have their greatest spectral response in the ultraviolet including compounds with carbonyl, nitro, nitrate, sulphydryl, alkyl iodides, alkyl bromides, acid chloride, thio carbonyl, and azo chemical functional groups (Hatayama 1986).

The sensor data are held in digital form and manipulated by computer. After calibrating the system on specific materials on the ground, it is then possible to instruct the computer to identify all other sites with similar spectra. The resulting map is displayed on a visual display unit, from which it can be turned into a photographic map. Such a system has been used to map specific forms of waste materials in the Lower Swansea Valley. It also offers considerable scope for monitoring the progress of restoration (Coulson and Bridges 1984, 1986). Ground-penetrating radar can be used to identify buried objects, such as buried drums of chemical waste, to detect plumes of contaminated leachate, and to identify the interface between disturbed and undisturbed soils. In addition, scans of the thermal emissivity of derelict sites can identify subterranean combustion and distinguish materials by their different heat-emitting characteristics.

From a photogrammetric survey of derelict land it would be possible to obtain precise site boundaries; accurate contours of the site; outlines of ruined buildings, waste heaps, ponds, or lagoons on site; and an up-to-date photomosaic of the derelict area and its surroundings.

Geological examination

Structural It is essential to know the geology below a derelict site before planning its reclamation. The geologist examines the rock outcrops in the derelict area and its neighbourhood in conjunction with the published geological map and attempts to construct a three-dimensional picture of the solid strata and overlying unconsolidated materials at the site. The geologist will also attempt to predict the depth of unconsolidated material since the presence of solid rock greatly increases the cost of landscaping. Not all geological materials are suitable for restorative work so the geologist should indicate in his report materials which are potentially suitable for subsoils and shall be placed on one side for later use, and unsuitable materials which should be completely buried.

Hydrogeological survey If toxic materials are to be dumped on the site as part of the reclamation of derelict land, or if the site has already been used for the disposal of toxic materials, it will be necessary to bring in a hydrogeologist. The hydrogeologist will investigate underground geological structures in the light of surface- and ground-water hydrology and will advise upon the possible movement of liquids and potentially dangerous leachates from the restored site. There are two alternatives—either to concentrate and contain any toxic materials, or to dilute and disperse them. Earlier practice tended to favour the first. Recently, it has been found that rock materials in the unsaturated zone below many landfill sites can be extremely effective at attenuating the concentration of toxic leachates and so it is not always necessary to locate watertight sites (Aspinwall 1975, 1976; John and Davies 1976; Mather and Bromley 1976; (Sumner Report 1978). Where ground water is immediately below a landfill site, the attenuation of leachates will not be so effective and toxic materials may move as a plume of contamination from the site in the direction of the ground-water flow (Fig. 5.5).

Geophysical survey In principle, geophysical equipment can detect boundaries between waste materials, layers of underlying rock, saturated or freely drained rock, cavities, and anomalies caused by buried objects and discontinuities. In practice, their significance is often difficult to interpret. Techniques include:

Seismic reflection—an induced seismic shock caused by striking a plate lying on the surface, or by exploding a small charge, is reflected by interfaces between layers of materials with different density. Geophones on the soil surface record the initial and reflected shock waves from buried objects which are then interpreted by the geophysicist.

Resistivity—an electric current is passed between two electrodes placed in the soil. The resistance to the current of electricity is measured at different intervals of separation. The results can be interpreted to detect the interface of different layers and the presence of cavities such as old mine shafts. Groundwater and dissolved chemicals affect resistivity.

Gravity survey—the presence of mineshafts, and buried quarry walls can be detected by measuring minute changes in the gravitational pull. Highly sensitive apparatus is available, but in the drastically disturbed conditions of derelict land it is difficult to identify problems with confidence.

Electromagnetic induction—an induced electromagnetic field at the soil surface generates a secondary field in any buried metallic object. The difference between primary and secondary magnetic fields can be measured and interpreted to identify buried objects. Unfortunately, most derelict sites are so

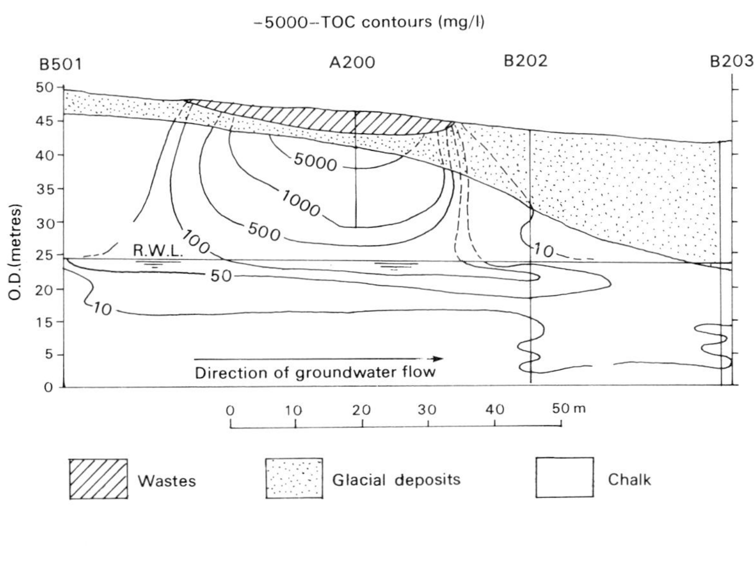

Fig. 5.5. Attenuation and plume of contaminants from a landfill site. (Crown Copyright, reproduced by permission of the Department of the Environment, from Sumner Report 1978.) TOC, Total organic content; RWL, Regional water-table level.

complex that the operator receives very confused information. However, single objects can be located by this method (Reinhold *et al.* 1986).

Magnetometer survey—buried metallic objects, mineshafts, and boundaries between different types of fill or rock type can cause local changes in the earth's magnetic field, but small features may be difficult to interpret with confidence.

In general, a geological survey of derelict land should provide information on the underlying rocks, their structures, and the nature and properties of superficial materials; the presence of geological material which could be used in the process of restoration; the presence of any shallow rock strata which may interfere with landscaping; the relationship of the site to local water sources, springs, and seepages; the presence of any metallic objects, voids, or impenetrable layers beneath the site.

Physical survey

The physical state of soils and waste materials lying upon the landscape determines site stability and whether plants will be able to establish themselves and maintain a cover. The physical attributes normally measured include the particle size of the materials, bulk density, particle density, porosity, infiltration rate, water-holding capacity, pore water pressure, liquid and plastic limits. In addition, thermal and radioactivity surveys may be required in some cases (Brink *et al.* 1982).

Many physical properties of soil and soil-like materials are related to their particle size and moisture content. On land disturbed by mining activities the soil frequently has a wide range of particle size, but often lacks the clay-sized material which is so important in natural soils. Many metalliferous slags were molten when tipped and now form large fused lumps of slag. Lagoons are often constructed on waste ground to receive finer material.

When a derelict site is disturbed, the dust produced can be an irritant to those who live or work nearby. Former smelters working metals, such as copper, lead, or zinc, have produced particulate matter which can be blown by the wind to pollute the land for many kilometres around (Welsh Office 1975). The 'moss-bag' technique of Goodman *et al.* (1975) is a method of sampling this dust.

In the field, the particle-size class of a soil or soil-like material may be determined manually on a moist sample. Usually, some samples are taken to the laboratory to check the particle-size class by sieving and sedimentation (Avery and Bascomb 1974). Analysis by sedimentation techniques does not work satisfactorily for materials such as finely divided coal and pulverized fuel ash.

Materials on derelict lands show a wide range of bulk density. Mine waste deposited from an overhead ropeway or at the top of a conical tip is not compacted and has a low density, whereas material sedimented from water may be closely packed and quarry floors may become tightly compacted by traffic. Certain blast furnace slags and pulverized fuel ash have pozzolanic properties which cause them to cement.

A physical survey of derelict land should include information about the particle size of the materials on the site; the stability of materials, especially on slopes, under normal and abnormal conditions of loading; an indication of the moisture limits beween which soils and waste material may be worked; the presence of combustible materials and radioactive materials.

Thermal and combustibility surveys Subsurface combustion affects many colliery waste tips and may occur in tips of domestic refuse and ash dumps. As the fire may lie beneath a thin, fragile crust, it presents a direct hazard to personnel and also emits asphyxiating or poisonous fumes and dust. Investigation of such sites may be dangerous and should never be carried out alone. Planks should be used for support and a harness with ropes worn in case of collapse; breathing apparatus may be required. In some cases it may be possible to consolidate the burning material with a heavy weight dropped from a dragline.

Thermocouples are more robust and reliable than mercury-glass thermometers in identifying hot areas. They may be lowered down holes or driven into the ground with probes and in some cases, thermal and infra-red imagery from remote sensors can be used (see p. 76). If the combustibility of a material is thought to be a potential problem samples should be taken for a calorific test. Large samples of several hundred kilograms are required.

Radioactivity surveys Testing and sampling for radioactivity is a task for suitably qualified personnel. The preliminary site survey should show whether the site must be surveyed for radioactive substances.

Typically, an on-site testing programme would include a gamma ray survey at different heights above the surface; alpha, beta, and soft gamma test counts on samples from the surface, trial pits, and within boreholes. If this shows that there is radioactive material on the site, it should be handled by suitably protected, experienced staff according to the provisions of the Ionizing Radiations Regulations 1985. Radioactive wastes may not be removed from a site until authorization has been received from the Radiochemical Inspectorate.

Soil surveys

There will be natural soil at most derelict sites, particularly those in country districts. Unless this is contaminated, it will be a valuable asset in the restoration.

A map of soils on a site will assist calculations of the quantities of material available for soil replacement after other earth-moving has been completed. The soil map must indicate the depths at which the contractor should separate topsoil and subsoil for storage. In appropriate cases subsoil material may be used where topsoil is in limited supply. In a recent soil survey prior to the opening of an opencast site in Dyfed four categories of material were recognized: (a) original topsoil; (b) potential topsoil derived from better quality subsoil; (c) potentially permeable subsoils derived from loose stony material; (d) impermeable subsoil material of inferior soil-forming quality (Wright 1981).

Usually, measurements of clay content, liquid and plastic limits, bulk density, water-holding capacity, and the depths of horizons are reliably related to soil mapping units. Thus, a map of the soils present assists the site engineer to decide how to remove the soils and how to avoid damage to soils during removal and restoration.

A map of soils can also be reinterpreted as a land-use capability map indicating the main limitations of the soils present for agriculture. This is particularly useful for comparing the productivity of land before disturbance and after restoration of opencast sites.

A soil survey during the assessment stage of a restoration programme should normally: indicate the extent of identified soil-mapping units by a detailed site plan; specify the thickness of topsoil and subsoil with calculations to indicate the *in situ* volume of the mapped soils, with information on their physical, chemical, and biological properties which may affect handling and storage properties; identify any soil parent material suitable as a substitute subsoil to remedy any deficiency of topsoil and subsoil.

Chemical surveys

The normal methods of soil analysis, described in textbooks such as Hesse (1971), Jackson (1958), or handbooks (Avery and Bascomb 1974), may be useful for the analysis of materials present on derelict sites. Analysis of plant material is described by Allen *et al.* (1974). However, it is important to bear in mind that derelict land frequently contains a variety of materials which are unlike natural soils in their chemical and physical characteristics. Thus, the interpretation of results, when obtained, has to be undertaken with care.

Waste materials lying upon derelict land are often pseudostratified, reflecting the manner in which they were dumped. As time passes, the surface of tipped materials becomes weathered and some elements are released from the surface layers and carried deeper into the waste heap, enriching the lower layers in a similar manner to leaching and eluviation in natural soils (Chase and Wainwright 1983).

Natural soils within and surrounding a derelict area frequently will be contaminated with dusts and drainage waters from toxic materials. These can be rated by soil and atmospheric surveys (Welsh Office 1975). Once in the soil metallic elements act in different ways. Lead is strongly held in the surface horizons associated with organic matter. Copper, cadmium, zinc, and nickel all show a tendency to accumulate in the lower soil horizons (Chase 1978; Bridges *et al.* 1981).

It is necessary to design an analytical programme which will take these variations into account. Decisions about which elements should be determined will be influenced by the previous site history. In most situations it is preferable to analyse for a limited number of important contaminants on a large number of samples, rather than undertake a fully comprehensive laboratory examination of only a few.

The determinations most commonly attempted on samples from derelict and contaminated land include:

pH; salinity; acid-soluble sulphate; elemental sulphur; sulphide; cyanide; ferrocyanide; ferricyanide; thiocyanate, free cyanide (where total cyanide is >20 ppm); phenols; toluene extract; cyclohexane extract; coal tar (where toluene or cyclohexane extract is >5000 ppm); organic chlorine; mineral oil.

Total amounts of lead; cadmium; mercury; antimony; arsenic; manganese; chromium; selenium; barium; vanadium; beryllium.

Plant-available lead; zinc; nickel; copper; boron.
Asbestos fibre content.
Combustibility.

Methods of analysis may require modification when applied to materials from derelict or contaminated land, and individual analysts have their own laboratory practices (Pickering 1977). Particular care should be taken to ensure that the laboratory knows why certain analyses have been requested; equally when the results of analyses become available for discussion, it is important that the environmental scientist is made aware of the method of analysis. This is of critical concern when dealing with 'total' and 'available' amount of contaminants and plant nutrients.

Chemical surveys of waste materials and soils of derelict areas will inevitably be expected to answer the question whether or not plants will grow in these materials when the site is reclaimed. Therefore, in addition to the list of analyses given previously, the major plant nutrients are also of concern in a survey of derelict or contaminated land. Available nitrogen is usually low or very low in amount in waste materials and soils of derelict lands and severely limits the growth of all plants. Some 'fossilized' nitrogen may be present in colliery shale, but for practical purposes this can be discounted, as only small amounts are available for plants (see p. 99). Phosphates and potash are also likely to be insufficient for satisfactory plant growth, but potash is usually adequate in coal-shale material. Trace elements are necessary for adequate plant growth, and most soils possess adequate levels although specific elements may be lacking or present in toxic amounts. Industrial activity is usually responsible for adding to the levels of trace elements which can exceed recommended safety levels.

Russell (1973) indicated that small amounts of iron, manganese, zinc, copper, boron and molybdenum are essential for normal plant growth and ruminant animals require cobalt. An indication of the normal levels of elements and the range over which they occur is given in Table 5.2. Similar background figures are given for Dutch soils by Edelman and de Bruin (1986).

A chemical survey of a derelict and contaminated site should indicate: the presence of any element or compound of a toxic or hazardous nature; the distribution and concentration of such elements or compounds on the site; the presence of any materials which might be used to neutralize each other, or which together might present an increased level of hazard; methods of dealing with any hazardous chemicals found on site; the presence of any benign materials which might be used to cover or isolate hazardous materials; the level of plant nutrients, making any necessary recommendations to facilitate plant growth.

Ecological survey

Ecologists have undertaken many studies of derelict land specifically with the aim of trying to get plants to grow upon and cover the unattractiveness of dereliction. Even when a restored site is to be used predominantly for buildings, it is still pleasant to have ornamental gardens and trees for amenity purposes (Spellerberg 1981).

Industrial waste materials have provided the ecologist with interesting and valuable habitats for study (Chadwick and Goodman 1975; Cairns *et al.* 1977; Davis and Beckett 1978). Each has a distinctive character, its approximate age is usually known, and its ecology is relatively simple. The effects of toxic materials upon plant growth can be assesed upon these sites. It has been found that the most successful plants are those which require low levels of nitrogen and phosphate, such as the grasses *Agrostis tenuis, Festuca rubra,* and *Festuca ovina.* Nitrogen fixers which can withstand low levels of phosphate include *Lotus corniculatus* and *Trifolium dubium.*

One of the first stages in any ecological survey of derelict land should be a field evaluation of the site. The plants which have voluntarily colonized a derelict site will possess inherent advantages which can be exploited in a reclamation scheme. These species also act as indicators of the ecological potential of the area. Many derelict lands are contaminated with waste materials which are either strongly acidic or strongly alkaline and can be tolerated only by calcifuge and calcicole species respectively. The range of pH values over which certain common British grasses occur is given in Table 5.3 and the associations of common flowering plants which can be found on acid, neutral, saline, waterlogged, and calcareous habitats of the United Kingdom are given in Table 5.4. It must be appreciated that the number of plant species present on a site is dependent upon many factors other than edaphic ones. These include the close proximity of natural vegetation to produce seeds and the physical conditions may limit the establishment of plants through unreliable moisture supplies. A rich plant cover with many species normally indicates a

Table 5.2. Elemental composition of soils (mg/kg)

Element	Median soil content (range)	Element	Median soil content (range)
Ag	0.05 (0.01–8)	Mg	5000 (400–9000)
Al	71 000 (10 k–300 k)	Mn	1000 (20–10 k)
As	6 (0.1–40)	Mo	1.2 (0.1–40)
Au	0.001–0.02?	N	2000 (200–5000)
B	20 (2–270)	Na	5000 (400–9000)
Ba	500 (100–3000)	Nb	10 (6–300)
Be	0.3 (0.01–40)	Nd	35 (4–63)
Bi	0.2 (0.1–13)	Ni	50 (2–750)
Br	10 (1–110)	O	490 000
C (org.)	20 000 (7 k–500 k)	P	800 (35–5300)
Ca	15 000 (700–500 k)	Pb	35* (2–300)
Cd	0.35 (0.01–2)	Pr	7 (3–12)
Ce	50 (3–170)	Ra	0.8 μ (0.2 μ–5 μ)
Cl	100 (8–1800)	Rb	150 (20–1000)
Co	8 (0.05–65)	S	700 (30–1600)
Cr	70 (5–1500)	Sb	1 (0.2–10)
Cs	4 (0.3–20)	Sc	7 (0.5–55)
Cu	30 (2–250)	Se	0.4 (0.1–2)
Dy	5 (2–12)	Si	330 000 (250 k–410 k)
Er	2 (0.6–6)	Sm	4.5 (0.6–23)
Eu	1 (0.1–3.2)	Sn	4 (1–200)
F	200 (20–700)	Sr	250 (4–2000)
Fe	40 000 (2 k–550 k)	Ta	2 (0.4–6)
Ga	20 (2–100)	Tb	0.7 (0.1–1.6)
Gd	4 (2–6)	Th	9 (1–35)
Ge	1 (0.1–50)	Ti	5000 (150–25 k)
Hf	6 (0.5–34)	Tl	0.2 (0.1–0.8)
Hg	0.06 (0.01–0.5)	Tm	0.6 (0.3–1.2)
Ho	0.6 (0.4–2)	U	2 (0.7–9)
I	5 (0.1–2.5)	V	90 (3–500)
In	1 (0.7–3)?	W	1.5 (0.5–83)
K	14 000 (80–37 k)	Y	40 (10–250)
La	40 (2–180)	Yb	3 (0.04–12)
Li	25 (3–350)	Zn	90 (1–900)
Lu	0.4 (0.1–0.7)	Zr	400 (60–2000)

*12 prior to global contamination; up to 100 in urban soils (Patterson 1979).
(Reproduced by permission of Academic Press and the author, from Bowen 1979.)

reasonable potential for restoration and a poor cover with few species suggests toxicity, nutrient deficiency, or physical problems.

If the ecologist is given time to complete exhaustive studies of the growth potential of a particular site and the materials upon it, his work might take place in three phases–a field survey, followed by controlled experiments in pots, and finally extensive experiments on a field scale.

An ecological survey should indicate: the plant communities present and the range of individual species colonizing the derelict site; plants which would be successful in reclamation and the preparations necessary to ensure satisfactory growth; which site materials should be placed out of the reach of plant roots, and which can be safely used as a growing medium; and, if the site is to be reclaimed and used for industrial purposes at some future date, a suitable temporary cover of plants.

Quantity survey

The role of the quantity surveyor is to estimate accurately the amounts of materials and the costs involved in moving these in a restoration. Although traditional methods will continue to play a central role in quantity survey, members of the reclamation team of the Durham County Council, working on the site of the Consett Steelworks, found that computer-generated models gave many advantages as they enabled the generation of cross-sections,

Table 5.3. Grasses particularly suitable for derelict land in temperate regions

Grass species	Fertility demand[1]	pH tolerance[2]	Drought tolerance[1]	Temperature tolerance[3]	Cultivars
Agrostis gigantea (red top)	M	NC	M	CW	Few
Agrostis stolonifera (creeping bent grass)	M	NC	L	CW	Few
Agrostis tenuis [bent grass (brown top)]	L	AN	M	CW	Many
Alopercurus pratensis (meadow foxtail)	M	NC	M	CW	Few
Dactylis glomerata [cocksfoot (orchard grass)]	H	NC	M	CW	Many
Deschampsia caespitosa (tufted hair grass)	M	NC	M	CW	Wild species
Deschampsia flexuosa (wavy hair grass)	L	A	H	CW	Wild species
Festuca arundinacea (tall fescue)	H	NC	L	CW	Many
Festuca ovina (sheep's fescue)	L	ANC	H	CdCW	Few
Festuca rubra (red fescue)	L	NC	H	CdCW	Many
Festuca pratensis (meadow fescue)	M	NC	M	CW	Few
Lolium perenne (perennial ryegrass)	M	NC	M	CW	Many
Phleum pratense (timothy)	M	NC	M	CW	Many
Poa compressa (Canada bluegrass)	L	ANC	H	CdCW	Many
Poa pratensis [smooth-stalked meadow grass (Kentucky bluegrass)]	L	ANC	H	CdCW	Many

[1] L = low; M = medium; H = high.
[2] A = acidic; N = neutral; C = calcareous.
[3] Cd = cold; C = cool; W = warm.
(Reproduced by permission of Blackwell Scientific Publications, from Bradshaw and Chadwick 1980.)

existing relief models, proposed relief models, and photomontages in minimal time (Connolly and Veitch 1983).

The quantity surveyor normally works from engineering drawings and specifications to produce a bill of quantities which forms the basis of the contract between the client and the contractor. The bill of quantities aims to provide a complete list of all materials and labour necessary for the completion of an engineering project so that an estimate of the costs can be made (Johnson and King 1947).

Assessment of data

When the findings of the site survey and the laboratory analyses are brought together, it is possible to begin the process of assessment. The aim of the surveys was to determine the nature of materials and contaminants on a derelict site so the hazards present may be assessed. Once the hazards are known, it is possible to relate them to potential 'targets'—adults, children, plants, animals, and buildings. The reclamation should be designed to minimize or avoid contact with actual or potential danger (ICRCL 1983). Hazards may exist through:

1. **Direct contact**—for example, substances which occur on old gasworks sites, such as phenols and thiocyanates and many other chemicals, are skin irritants and possibly carcinogens. The presence of these substances requires protection for workers during restoration and perhaps subsequently for workers who excavate into the ground.
2. **Ingestion**—direct ingestion of poisonous

Table 5.4. Flowering plants commonly associated with particular habitats in the United Kingdom

Acidic	**Calcareous**
Agrostis canina (brown bent)	*Achillea millefolium* (yarrow)
Agrostic tenuis (bent grass)	*Brachypodium pinnatum* (torgrass)
Betula pendula (birch)	*Briza media* (quaking grass)
Deschampsia flexuosa (wavy hair grass)	*Festuca rubra* (red fescue)
Digitalis purpurea (foxglove)	*Fraxinus excelsior* (ash)
Galium saxatile (heath bedstraw)	*Helictotrichon pratensis* (oat grass)
Pteridium aquilinum (bracken)	*Trisetum flavescens* (yellow oat grass)
Rumex acetosella (sheep's sorrel)	
Ulex europaeus (gorse)	**Waterlogged**
Vacinium myrtillus (bilberry)	*Agrostis stolonifera* (creeping bent)
	Alnus glutinosa (alder)
Neutral	*Caltha palustris* (marsh marigold)
Agropyron repens (couch grass)	*Deschampsia caespitosa* (tufted hair grass)
Aira caryophyllea (silver hair grass)	*Filipendula ulmaria* (meadow sweet)
Anisantha sterilis (sterile brome)	*Juncus articulatus* (jointed rush)
Arrhenatherum elatuis (false oat)	*Juncus effusus* (common rush)
Bromus mollis (soft brome)	*Phalaris arundinacea* (reedgrass)
Atriplex hastata (orache)	*Phragmites australis* (reed)
Chamaenerion angustifolium (willow herb)	*Salix atrocinerea* (sallow)
Cirsium arvense (creeping thistle)	*Salix fragilis* (crack willow)
Cirsium vulgare (spear thistle)	*Typha angustifolia* (reed mace)
Hordeum murinum (wild barley)	*Valeriana dioica* (marsh valerian)
Poa annua (annual meadow grass)	
Poa pratensis (smooth-stalked meadow grass)	**Heavy metals**
Tussilago farfara (coltsfoot)	*Agrostis tenuis* (bent grass)
	Agrostis stolonifera (creeping bent)
Saline	*Festuca ovina* (sheep's fescue)
Agropyron junceum (sea couch)	*Festuca rubra* (red fescue)
Atriplex littoralis (orache)	*Minuartia verna* (spring sandwort)
Cakile maritima (sea rocket)	*Silene vulgaris* (bladder campion)
Festuca rubra (red fescue)	
Hippophae rhamnoides (sea buckthorn)	
Honckenya peploides (sea purslane)	
Puccinellia maritima (sea poa)	

(Reproduced by permission of Blackwell Scientific Publications, from Bradshaw and Chadwick 1980.)

substances, particularly by children with dirty fingers; cyanide and elements such as lead, cadmium, arsenic, selenium, antimony, chromium, mercury, and any others which the site history might suggest. Indirect ingestion could occur through the consumption of vegetables grown in contaminated ground, and animals could similarly suffer through ingestion of elevated metal levels.

3. **Inhalation**—inhalation of gases can occur from materials on derelict sites, particularly where water levels have been changed. Dusts containing heavy metals and asbestos are hazardous during reclamation and could remain so, unless well buried.

4. **Flammability**—coal wharves, rail sidings, and old timber yards may contain large quantitites of combustible material. Old gasworks sites can have materials with up to 10 per cent sulphur.

5. **Structural**—concrete structures and plastic pipes can be attacked underground by water-soluble sulphates and organic solvents, respectively. Iron pipes are corroded more rapidly in strongly acidic conditions.

The ICRCL (1983) guidelines refer to the level of contaminants present in surface materials of a contaminated or derelict site. These figures refer to 'spot samples', individual unbulked samples from specific locations in the sampling pattern. If any sample exceeds the 'trigger concentration' then the whole area should be regarded with suspicion, but concentrations exceeding the guidelines are not necessarily unacceptable; they may only be undesirable for certain land uses. It is suggested that if more than 1 per cent of the land surveyed exceeds the guidelines then the level of contaminant should be regarded as unacceptable.

A concept of the 'worst case concentration' has been suggested by JURUE (1982). This is described as '. . . the contamination level exceeded by the worst-contaminated 2.5 per cent of the land within a radius of around 1 km of any given location'. Using a statistical approach, this attempts to overcome the high spatial variability of surface materials on derelict land. Averaging gives an unhelpful impression of contaminant levels, but, if a significant fraction of the area exceeds the guideline figures, this is helpful. In a survey of Walsall, 'hotspots' of contaminants have been highlighted by calculating 'moving four square populations' using the measurements from four adjacent 1 km^2 sampling areas. Although only seven samples were taken per km^2, it is possible to calculate what the situation would be if it had been possible to take 40 samples; one of these (2.5 per cent) could be expected to have levels as high as the worst case concentration. (In this JURUE survey, 2.5 per cent was used rather than the 1 per cent recommended by ICRCL (1983) because of the sampling procedures.)

The impact of specific hazards upon the use of a site enables contaminants to be ranked according to their severity. Some substances, such as asbestos, or sites which are producing methane are of sufficient danger or difficulty to set the trigger concentration at zero. A second grouping includes several contaminants which can be tolerated in small amounts in 'soft' developments but which would be too dangerous where close contact with the ground is likely, as in housing development. There is a third group of contaminants which does not have any clear relationship between the level in the environment and human health. The trigger values for these contaminants are difficult to determine as insufficient is known about the immediate or long-term effects. The contaminants and their trigger concentrations given in this account are listed in the current *Guidance on developing contaminated land* (ICRCL 1983) and shown in Table 4.2, p. 40; they will be updated from time to time as new information becomes available. Comparable figures used in the Netherlands are given in Table 4.1, pp. 38–9.

The use of the trigger concentrations is based on a number of assumptions which must be observed when a site assessment is being made. All the values refer to the concentration of contaminants in 'spot' samples obtained from an adequate site survey. They do not apply to analyses of samples which have been averaged, bulked, or are composite. If all the sample values lie below the trigger concentrations, the site may be regarded as uncontaminated as far as the hazards from these contaminants are concerned and development may proceed. Remedial action may be necessary when the levels of contaminants exceed the trigger concentrations, but this will be dependent upon the proposed land use and whether the contaminant is still being added to the site by fall-out or seepage.

Such assessments may be helpful in the planning of prospective developments and the reclamation which precedes them. In order to meet conditions of the Health and Safety at Work Act, a developer will need to know in advance about dangerous materials on site. There are no official or statutory limits published for contaminants present on derelict or contaminated land (ICRCL 1983). However, the concept of 'trigger concentrations' below which land may be considered to be uncontaminated is useful. These concentrations are intended for guidance and are not meant to distinguish sites where remedial action is necessary, nor do they represent maximum permissable concentrations. The decision on what action to take in individual cases rests upon the proposed future land use and the impact the contamination will have on the users or future occupants.

Contractural procedures for land reclamation

Drafting the contractural documents takes place according to accepted working practices of tendering and contract between the client and the contractor. It is recommended that the Institute of Civil Engineers Conditions of Contract and Standard Methods of Measurement are used, even if they are not perfectly suited to derelict land reclamation problems. The Institute of Landscape Architects also has an approved form of contract. The aim of such documents is to be legally correct, comprehensive, and free from any ambiguities should any problems arise during the restoration programme.

Tender documents usually include the site plans, details of the contractual conditions, and the bill of quantities or specifications. This information should be sufficiently comprehensive to enable the tendering contractor to make his own estimates of costs and requirements to complete the programme. Clark (1977) recommends that tendering and contractual instructions should be included in the same document. Tenders may be sought on the open market or invited from a selective list of known and

CONTRACTS & TENDERS

Greater Manchester Council and Lancashire County Council

JOINT RECLAMATION TEAM

RECLAMATION OF DERELICT LAND

FORMER SHELL/ICI SITE, HEYSHAM, LANCASTER

SELECTIVE TENDERING

Applications are invited for a select list of tenderers for the Reclamation Scheme at the former Shell/ICI site, Heysham, which involves the following works:

(i) General clearance of the 77ha site including the filling of redundant tanks, manholes, sewers and pipe ducts.
(ii) The provision of a surface water drainage system involving approx. 2,000m of pipe up to 900mm diameter and 3,000m of open ditch.
(iii) General excavation and reshaping of the site including filling in compacted layers approx. 300,000m3.
(iv) The construction of flexible pavement roads to serve proposed industrial areas including improvement of existing roadways approx. 15,000m2.
(v) Infrastructure associated with industrial development including 1,700m of foul sewer provision of water and electricity and street lighting.
(vi) Landscaping cultivation and seeding of certain areas of the site.
(vii) Fencing to agricultural areas and to secure industrial areas.

Civil Engineering Contractors who have experience of this type of work are invited to apply for inclusion on the list, giving particulars of similar contracts they have recently undertaken, including the location and the employing authority, together with the name and address of the Engineer for each project, who will be willing to testify as to the applicant's experience. A copy of the latest Profit and Loss Account and Balance Sheet should be returned with the application to The Principal Engineer, Joint Reclamation Team, Westward House, King Street West, Wigan WN1 1LP by Thursday, October 28, 1982.

The Council does not bind itself to invite Tenders from all or any of the applicants, and no correspondence will be entered into concerning the Select List.

Fig. 5.6. Newspaper advertisement for tenders. (From the *Guardian*, to whom thanks are due.)

competent contractors. An example of an advertisement used is given in Fig. 5.6.

Many of the tasks of restoration require specialist services, so the form of contract normally specifies that subcontractors must be approved by the engineer in charge. Normally, the contract will also include a surety and guarantee bond as a safeguard for the completion of the work. The contractor whose tender is accepted will be expected to commence work as soon as practicable after the starting date of the contract but some flexibility is allowed because of possible adverse weather conditions.

The specification and bills of quantities describe the various elements of the works required, the standard of workmanship, and the amounts of materials involved. Conventional specification documents require significant changes to be relevant to derelict land, so authorities with extensive areas of dereliction have evolved their own form of specification. The following points should be covered:

Overheads estimates. These include contractors' site buildings, supervision of the site, welfare provision for operators, and insurance.

Demolition and site clearance estimates. Clean rubble has a value as hardcore and if not required to make up ground levels can be sold off the site. Foundations and cellars are a problem because of the unevenness of the undisturbed ground beneath. Concrete slabs and foundations can normally be cleared with light machinery but the foundations and bases of high chimneys and furnaces could be very difficult to remove especially where preliminary demolition has covered them with debris. In coalfield areas, old mine and air shafts occur which must be filled and capped to avoid subsidence.

Earthworks. It is recommended that considerable care be given to calculations and planning of earthworks as these can absorb more than half of the costs of land reclamation. As has been expressed previously, knowledge of the future land use can help with decisions about the new land form which is to be constructed. Toxic materials may have to be completely removed if housing is contemplated; industry requires large level areas; agriculture, similarly, requires large areas with low angles of slope, but can be interspersed with forestry which can accept steeper slopes. In all these cases, the amounts of materials involved in earth-moving are critical and must be accurately estimated in advance. The cost of stripping topsoil and subsoil in relatively thin layers, to be separately banked, must be calculated, as must the greater expense of obtaining and transporting uncontaminated soil to the site. Depth of topsoil and subsoil to be removed should be stipulated.

Drainage estimates. In the western part of the British Isles, excess soil water is a problem which must be taken into account when restoration is considered. During restoration it is a requirement that there is no sediment-laden water leaving the site which may pollute natural water courses. It will be necessary in many cases to construct sediment ponds to prevent this from happening. Each site will require a specific plan for the safe disposal of water, minimizing the effects of sheet and gully erosion with low gradients and stone-lined sections where steep falls occur. Culverts beneath roads and pathways should be included in the calculations. Subsoil and topsoil should be replaced such that drainage can be installed. Some temporary contour-ditches may be required initially and will require maintenance during the establishment period, although they will be filled in ultimately.

Difficult handling conditions. Estimates should include the additional costs involved when difficulties in handling are present. If haulage is contemplated this can add to costs, especially on hilly terrain. Where burning or hot materials, abrasive materials, or slurries are involved, wear and tear on machinery is greater, increasing the contractor's costs. The bill of quantities should specify the type and amount of materials to be moved and the methods of hauling, deposition and compaction to be employed in the restorations.

Contaminants. Where contaminants exist and an earth-moving operation is to take place, the design of the restoration work should incorporate the engineering work as a means of removing or reducing contamination (see Chapter 6). The contaminated material can be dealt with in one of three ways:

1. It can be neutralized *in situ* by chemical means, or the physical environment may be changed as with lowering the water-table or improving the site drainage to render the contaminant harmless.
2. The contaminated material can be encapsulated or isolated by natural materials such as clay and covered by an impermeable capping followed by clean soil material. Isolation beneath concrete or asphalt as in a car park is often the most straightforward solution.
3. The offending material may be completely removed and replaced by clean material. This solution has the problems of extra expense and the difficulty of disposing of the material somewhere else. There is also the danger of spreading the pollution along the haulage route.

Landscaping, cultivation, and maintenance estimates. The slope and size of fields may be dictated by the lie of the land following restoration, but it is strongly recommended that these should conform with the surrounding undisturbed land. Allowance should be made for fences (thorn and post and wire), gates, stiles, and rabbit-proof netting for forestry areas. Farm roads should be constructed according to ministry specifications, preferably using materials such as burnt shale, found on site. On opencast restorations it is normal practice to place water

Table 5.5. Typical estimate for the reclamation and re-seeding of derelict land

		Unit cost (£)	Total cost (£)
Demolitions and site clearance			
Pull down and remove fences	4500 m	0.10	450.00
Pull down and remove building 29 × 8 × 4 m high complete with heavy concrete floor	Item		400.00
Ditto partly demolished brick structure 7 × 2 × 1 m high	Item		50.00
Ditto brick structure with concrete roof and floor slab 6 × 3 × 3 m high	Item		50.00
Ditto arched brick structure 5 × 2 × 1.5 m	Item		100.00
Ditto brick structure on heap 3.5 × 3 × 2.5 m high	Item		50.00
Break up remains of machine base 1 × 2 × 1 m	Item		20.00
Break up concrete slabs, roads and the like	800 m^2	1.00	800.00
Pull down and remove brick wall	90 m^2	0.50	45.00
Remove sundry debris and rubbish	Item		500.00
			£2465.00
Shaft caps			
Provision for investigating mine shafts and capping as required	3	700	2100.00
			£2100.00
Earthworks			
Excavate to remove topsoil and deposit in stockpiles	26 000 m^3	0.16	4 160.00
Excavate topsoil from stockpile and spread over forestry areas (av. 200 mm)	8 000 m^3	0.16	1 280.00
Excavate waste material and deposit on adjoining areas	422 000 m^3	0.12	50 640.00
Extra over last for removing slurry and burying in thin layers	40 000 m^3	0.41	16 400.00
Ditto burning material	3 000 m^2	0.31	930.00
			£73 410.00

Table 5.5. *Continued*

		Unit cost (£)	Total cost (£)
Drainage			
Permanent open ditch	1070 m	0.60	642.00
Temporary ditch back-filled during maintenance period	490 m	2.00	980.00
Ditch crossings	3	160.00	480.00
Catchpits and headwalls to ditches	2	160.00	320.00
750 mm outfall pipe to stream	60 m	30.00	1800.00
Headwall to last	1	200.00	200.00
Connection of catchpit to existing manhole in road	Item		400.00
			£4833.00
Water supply			
Excavate for and lay water main	220 m	0.50	110.00
Cattle troughs and stopcocks	2	55.00	110.00
			£220.00
Seeding and planting			
Initial cultivations, removal of stone, and grading	19 ha	80.00	1 520.00
Cultivate and seed shale areas	15 ha	450.00	6 750.00
(*As last*) new soiled areas for woodland planting	4 ha	400.00	1 600.00
Provision for two-year establishment period of grass on shale	15 ha	200.00	3 000.00
Woodland planting	9 ha	520.00	4 680.00
Provision for five-year establishment period for woodland planting	9 ha	260.00	2 340.00
			£19 890.00
Fences			
Post and wire fence	4890 m	1.50	7 335.00
Post and wire fence with netting	2035 m	2.00	4 070.00
Field gates	2	40.00	80.00
Field gates to woodland areas	3	45.00	135.00
			£11 620.00
Roads and paths			
Reinstate farm track 3 m wide	200 m	3.00	600.00
Bridle path in compacted red shale 1.2 m wide	100 m	0.15	15.00
			£615.00
Summary			
Demolitions and site clearance			2 465.00
Shaft caps			2 100.00
Earthworks			73 410.00
Drainage			4 822.00
Water supply			220.00
Seeding and planting			19 890.00
Fences			11 620.00
Roads and paths			615.00
Preliminaries 5 per cent on items 1–8			5 757.00
Contingencies 5 per cent on items 1–9			6 045.00
	Total estimated cost		£126 944.00

Note: The estimated cost of £126 944.00 does not include professional fees, land acquisition, legal and administration costs.

(Reproduced by permission of I. S. Clark and the University of Newcastle, from Clark 1977.)

pipes to cattle troughs in the fields. Such items can normally be budgeted by linear measure of fence road or piping required.

Once the physical work of replacement has been completed the site needs to be cultivated and the pioneer crop sown—usually a grass mixture—with appropriate lime and fertilizer treatment. Steep slopes may require special treatment such as hydro-seeding. Tree-planting is a labour-intensive operation and its cost varies according to the maturity of the trees. Maintenance during the establishment period can be an important charge and must be considered. The cost of soil sampling, analysis, and subsequent advice on levels of treatment must be included.

Supervision and certification estimates. After the tender has been accepted, supervision and cost control must continue throughout the restoration programme. Work has to be monitored on a day-to-day basis but there are also critical points where client and contractor must agree that quantities were correctly estimated, correct procedures followed, and the work satisfactorily completed. It may be necessary to have a site survey before final payments are made to ensure correct land form and the depth of subsoil and the topsoil upon it.

Table 5.5 contains a typical estimate for reclamation and re-seeding of derelict land.

Tables 5.6 and 5.7 give examples of the main framework for a specification and a bill of quantities

Table 5.6. Specification clauses

Description of site

Before commencement of cultivation and seeding the following initial cultivation and soiling operations will have been completed by the general contractor:

(a) Surface of colliery shale in all areas ripped to a depth of 600 mm (24 in) with a heavy subsoil rooter at 600 mm centres along the contours.

(b) All stones etc. of 150 mm (6 in) diameter or over removed and buried.

(c) Surface of shale graded and scrubbed.

(d) Subsoil spread to specified thickness on all areas except woodland.

(e) Subsoil ripped to its full depth, i.e. 300 mm (12 in).

(f) Stones etc. of 150 mm (6 in) diameter or over removed from surface of subsoil.

(g) Surface of subsoil graded or scrubbed.

(h) Topsoil spread to specified thickness on all areas except woodland.

(i) Stones etc. of 50 mm (2 in) diameter and over removed from surface of topsoil.

(j) Surface of topsoil graded or scrubbed.

Lime and fertilizers (provisional item)

The subcontractor shall allow the provisional sums included in the bills of quantities for the sampling and analysis of the soil and the supply of lime and fertilizers over the total area.

Before starting cultivation the subcontractor shall make arrangements for the sampling and analysis of the soil to determine the type and rate of application for lime and fertilizers and shall deliver and spread the lime and/or fertilizers over the site to the instructions of the engineer.

Cultivations—soiled areas

The subcontractor shall undertake the following operations over soiled areas in the order shown (average total depth of soils expected to be 400 mm; 16 in).

1. Spread ground limestone as specified.
2. Chisel plough 150 mm (6 in) deep (one pass).
3. Spread basic slag as specified.
4. Disc harrow (two passes).
5. Heavy harrow (one pass).
6. Remove all stones, wood, metal, and other deleterious material larger than 50 mm (2 in) diameter brought to the surface.
7. Scrub (one pass).
8. Distribute compound fertilizer as specified.
9. Light harrow (one pass).
10. Roll with Cambridge roller (one pass).
11. Sow grass seed as specified.
12. Light harrow (one pass).
13. Flat roll (one pass).

Cultivations—shale areas

1. Spread ground limestone as specified.
2. Chisel plough 150 mm (6 in) deep (one pass).
3. Spread basic slag as specified.
4. Disc (one pass).
5. Remove all stones, other deleterious materials larger than 50 mm (2 in) diameter.
6. Scrub (one pass).
7. Spread dried sewage sludge as specified.
8. Heavy harrow (one pass).
9. Distribute compound fertilizer as specified.
10. Roll with Cambridge roller (one pass).
11. Sow grass seeds mixture as the prescribed rate.
12. Light harrow (one pass).
13. Flat roll (one pass).

Grass seed mixtures

1. Grass seed mixture 'A'

30 kg	Perennial Ryegrass S23
14 kg	Cocksfoot S143
7 kg	Timothy S48
3 kg	Meadow Fescue S53
3 kg	Creeping Red Fescue S59
7 kg	White Clover S100
3 kg	Red Clover S123
67 kg/ha	(60 lb/acre)

This mixture shall be used on all pasture areas at the rate of 67 kg/ha (60 lb/acre).

Table 5.6. *Continued*

2. Grass seed mixture 'B'
   40 kg Red Fescue S59
   9 kg White Clover S184
   9 kg *Festuca tennuifolia* (certified strain)
   9 kg *Agrostis tenuis* (certified strain)
   67 kg/ha (60 lb/acre)

This mixture shall be used on all woodland areas at the rate of 67 kg/ha (60 lb/acre).

Cultivations and grass seeding—general

No lime, seed, fertilizer, basic slag, or dried sewage sludge shall be applied or cultivations carried out when ground or other conditions are unsuitable. The engineer's decision shall be final.

The prices of culitivations, treatments, and operations, where indicated in the foregoing, shall be based on the number of passes of the machine or cultivator specified. Additional passes, cultivations, or amendments to the specification will be paid for in accordance with the items marked 'provisional'.

Prices for the removal of stones etc. shall include the removal off-site, or burying not less than 900 mm (35 in) below the finished surface, of all materials of the size limit stated brought to the surface by these cultivations, notwithstanding that such materials conform to the description of materials to be removed during the initial cultivations.

No fertilizer or seed is to be sown until the engineer has approved the seed bed.

The subcontractor will be required to produce all delivery notes for lime, seed, basic slag, and compound granular fertilizers for inspection and recording.

The lime, fertilizer, and seed shall be supplied by a merchant approved by the engineer. The seed shall be mixed to the required specification at the merchant's works and delivered to site in sealed bags clearly marked with the code reference for the mixture.

The contractor must ensure that all machinery and implements used in establishing and maintaining the grass sward are properly guarded in accordance with the Agriculture (Safety, Health, and Welfare Provisions) Act 1956 and Regulations made thereunder.

Establishment period

The subcontractor shall maintain all the grassland for a period of 12 months commencing from the date of completion of the sowing of the grass seed and carry out the following work:

(a) Supply, deliver, and spread ground limestone and fertilizer as specified by the engineer and within two weeks of receiving written instruction (see provisional sum included in bills of quantities).

(b) Prior to the first cutting of the grass remove all stones, wood, metal, and other deleterious matter of 50 mm (2 in) diameter and over, and before completion of the establishment period remove all such similar materials that have worked to the surface. All such materials to be removed from the site.

(c) The subcontractor shall cut and mulch the grass when 100 mm (4 in) in height reducing the height to 50 mm (2 in). Avoid root-pulling and cutting in wet weather.

(d) The subcontractor shall throughout the remainder of the establishment period cut and mulch the seeded areas according to growth and shall allow for a cut to take place as often as the grass produces a further 100 mm (4 in) of growth from the previous cut and prevent the grass from seeding.

(e) It is anticipated that mowing will be required every 12 to 14 days during the height of the season but only once every 20 to 22 days at the start and finish. The sward should enter the dormancy period with a growth of no more than 100 mm (4 in) in height.

All cutting and mulching shall be undertaken using a Sapper Bamford B6 flail mower or Lupat Flailmaster or other machine to be approved by the engineer. The subcontractor will be required to prevent the mulched grass from collecting in rows or rills or in any other way deleterious to the growing sward.

During the establishment period the subcontractor shall direct his attention to establishing a good sward of this specific mixture free from pernicious and other weeds. In doing so the subcontractor shall apply approved weed-killer at the direction of the engineer.

respectively, for a 10 ha (25 acre) site reclaimed to provide 7 ha (17 acres) of agricultural land and 3 ha (8 acres) of woodland. The available soil and subsoil is spread on the agricultural area and the woodland area remains bare shale.

Contractual works

When the plans for reclamation have been agreed and contracts exchanged between the restoring authority and the contractor, the physical work upon the site can begin. This work is the responsibility of the site engineer and the role of the surveyor of derelict land is almost complete. However, he may become involved in the process of site supervision to ensure that the contractor and his men deal effectively and according to schedule, with the hazardous materials. Some of the methods of on-site, *in situ*, and covering systems are given in Chapter 6 where different types of restoration schemes are reviewed.

Monitoring

After the engineering work on site has been completed, it may be necessary to arrange for monitoring and/or maintenance to take place.

Table 5.7. Bills of quantities descriptions

Soil sampling and analysis	
Provide the provisional sum of £50.00 for soil sampling and analysis.	
Lime compound fertilizer and basic slag	
The type and rate of application of lime, fertilizer, and basic slag will depend largely upon the results of the soil analysis.	
All charges for taking delivery, unloading, storing, and transporting on site, profit, and all other charges in connection with the supply of limestone, compound fertilizer, and basic slag must be included in the items for the spreading of these materials. The prices for 'spread only' items shall therefore include all such charges.	
Provide the provisional sum of £1600.00 for the supply of lime compound fertilizer and basic slag for application during cultivation of the seed bed and establishment period.	
Cultivations and seeding—soiled areas	
Chisel plough, disc, heavy harrow, remove all stones, etc., scrub, light harrow, roll, light harrow, and roll as specified	7 ha
Spread only ground limestone at a rate of 12.50 tonnes/ha (5 tons/acre)	7 ha
Extra over last for a rate of 25.00 tonnes/ha (10 tons/acre) (provisional)	7 ha
Spread only basic slag at a rate of 0.63 tonnes/ha (0.25 tons/acre)	7 ha
Extra over last for a rate of 1.25 tonnes/ha (0.5 tons/acre) (provisional)	7 ha
Spread only compound granular fertilizer at a rate of 0.25 tonnes/ha (0.10 tons/acre)	7 ha
Extra over last for a rate of 0.50 tonnes/ha (0.20 tons/acre) (provisional)	7 ha
Supply and broadcast seed as specified at a rate of 67 kg/ha (60 lb/acre) (Mix A)	7 ha
Cultivations and seeding—shale areas	
Chisel plough, disc, remove all stones etc., scrub, heavy harrow, roll, light harrow, and flat roll	3 ha
Spread only limestone at a rate of 12.50 tonnes/ha (5 tons/acre)	3 ha
Extra over last for a rate of 25.00 tonnes/ha (10 tons/acre) (provisional)	3 ha
Spread only basic slag at a rate of 0.63 tonnes/ha (0.25 tons/acre)	3 ha
Extra over last for a rate of 1.25 tonnes/ha (0.50 tons/acre) (provisional)	3 ha
Supply, deliver, and spread dried sewage sludge at a rate of 5.00 tonnes/ha (2.0 tons/acre)	3 ha
Extra over last for a rate of 10.00 tonnes/ha (4 tons/acre) (provisional)	3 ha
Spread only compound granular fertilizer at a rate of 0.25 tonnes/ha (0.10 tons/acre)	3 ha
Extra over last for rate of 0.50 tonnes/ha (0.2 tons/acre) (provisional)	3 ha
Supply and broadcast seed as specified at a rate of 67 kg/ha (60 lb/acre) (Mix B)	3 ha
The following cultivation work all provisional	
Chisel plough (one pass—to a depth of 200 mm; 8 in)	3 ha
Disc (one pass)	3 ha
Heavy harrow (one pass)	3 ha
Light harrow (one pass)	3 ha
Roll (one pass)	3 ha
Establishment period	
Remove all stones, etc. and other deleterious matter 50 mm (2 in) diameter and over prior to first cut over soiled area	7 ha
(*As last*) but over shale areas	3 ha
Cut and mulch grass as specified over soiled areas (first cut)	7 ha
(*As last*) but additional cuts (5 No. provisionally)	35 ha
Cut and mulch grass as specified over shale areas (first cut)	3 ha
(*As last*) but additional cuts (5 No. provisionally)	15 ha

(Reproduced by permission of I. S. Clark and the University of Newcastle, from Clark 1977.)

Problem sites where it has been necessary to encapsulate or cover hazardous materials should be monitored to ensure there is no escape of contaminants. This may necessitate the taking of soil, water, and gas samples for many years after the cessation of the engineering work. In the long term, one of the greatest problems will be to ensure that contaminated sites remain undisturbed, unless there are overriding reasons why interference should occur.

A period of management is prescribed for restoration of opencast coal and ironstone areas. In an agricultural restoration, it is essential to ensure that the new soil can be ploughed and cultivated, that plants will grow, and that the drainage system is efficient. It is important, too, that the land is treated carefully, so that soil structure is redeveloped and the new soil ecosystem given every possible encouragement to reach stability. Although considerable experience has been gained from the management of these restored lands, little, if any experience has accumulated concerning the long-term stability of schemes where contaminated land has been reclaimed.

6 Design for restoration

Remedial measures should always attempt to solve the problem once and for all. Temporary, palliative measures are poor substitutes for permanent measures and usually the problem re-emerges. The final cost of restoration is then greater than if the problem had been dealt with effectively in the first place.

The simplest solution to the problem of derelict land is not to allow it to occur. For example, it has been official policy that areas disturbed by opencast mining for ironstone or coal should be restored for productive use. Unfortunately, similar provisions for restoration for sand and gravel workings were ineffective until recently. However, the position is much improved as a result of the activities of the Sand and Gravel Association, reinforced by the Stevens (1976) Report on the Planning and Control of Mineral Workings, and the Town and Country Planning (Minerals) Act 1981 (Fig. 6.1).

Earlier chapters have shown that derelict industrial land has many difficulties which need to be overcome before it can be used productively again. The Interdepartmental Committee on the Redevelopment of Contaminated Land of the DoE drew attention to the problems caused by contamination (ICRCL 1978*a*) and identified the presence of potential dangers to public health, as well as physical problems which add to the expense of clearance and redevelopment. If these hazards are not recognized during the site survey and adequate precautions taken, the consequences may be costly in terms of health and safety and very expensive to correct.

Fig. 6.1. Former gravel pit near Reading now used for recreational purposes.

Dealing with contamination

The systematic survey of derelict land described in Chapter 5 should have ascertained the character, concentration, and distribution of any contaminants on site. These must be rendered harmless before restoration can begin so their impact on subsequent land use is reduced. There are several possibilities—some well-tried, others still experimental. They will be discussed under the subheadings of on-site treatments, *in situ* treatments, and covering systems. They all have the advantage that contaminated material need not leave the site.

On-site treatments

Treatments such as leaching, separation, flotation, thermal decomposition, steam stripping, chemical, microbiological, and stabilization techniques were reviewed by Rulkens *et al.* (1985). They all involve excavation of the contaminated material, its treatment, and replacement after decontamination.

Leaching Contaminated soil is mixed with an extracting agent, usually in aqueous solution, to remove the contaminants. It is most convenient if the extracting agent can, in turn, be separated from the contaminants and reused. Hydrochloric, sulphuric, or nitric acids may be used to extract heavy metals; sodium hydroxide or carbonate may be used to extract cyanides of hydrocarbons. There may be problems from heavy metals being adsorbed on to clay and humus colloids, and humic materials may require large amounts of extractant. After extraction the soil must be cleansed of the extracting agent itself; this may be difficult and will usually leave the soil in a poor structural state.

Mechanical separation This has been used in the Netherlands on sandy materials where the specific gravity, particle size, and magnetic properties of the soil and the contaminant are different. If so, it may be possible to mix the soil with water and concentrate the contaminant into one size fraction, which can be separated by sieving or sedimentation.

Flotation If the contaminant occurs in silt-sized particles and their specific gravity or surface properties are different from the soil matrix, it may be possible to separate it by flotation as used in ore separation. This requires a suspension of the contaminated soil and water to which a chemical flotation agent is added, and the mixture violently stirred or bubbled so the contaminant may be removed with the resulting froth.

Thermal decomposition Volatile contaminants may be evaporated or decomposed by heating. This is usually accomplished in a rotary kiln followed by incineration in an afterburner. Dioxins or PCBs may require temperatures of up to 1700 °C for complete breakdown, but many organic substances decompose to harmless secondary products at lower temperatures, particularly if catalysts are used. The dusts produced during incineration must not be allowed to escape and vaporized chemicals are removed by gas scrubbers or absorbed on activated charcoal. The standard equipment for these processes is large and expensive to run, but mobile units have been developed in the United States and Germany so contaminants can be removed from soil actually on the site.

Steam stripping Soil contaminated with volatile compounds may be stripped by injection of steam into a rotating drum containing the soil. The volatile substances taken up by the steam are then condensed and separated from the water. Hydrocarbons, petrol, benzene, methanol, ethanol, kerosene, and halogenated hydrocarbons can be removed by this method.

Chemical treatments These include neutralization, hydrolysis, oxidation, and reduction. An intimate mixture of contaminated soil and the reagent is mixed into a slurry and a reaction allowed to take place. A lengthy contact time may be required and subsequent cleansing treatment to remove the resulting chemicals may be necessary. The decontaminated soil will be structureless. Modification of the pH by lime is frequently used to immobilize the heavy metals by lowering their solubility.

Microbiological treatment This is currently being explored but theoretically it should be possible to decompose cyanides or aromatics such as benzene and phenol, alkanes, cycloalkanes, and chlorinated hydrocarbons by bacterial degradation. Microbial treatments have successfully extracted metal from sulphide ores using strains of Thiobacillus. However, the process is slow and more appropriate for *in situ* treatments.

Stabilization A chemical reaction is sometimes employed to form insoluble or hydrophobic compounds which render the contaminants in the soil less soluble, or less obnoxious. Commercial stabilization techniques were developed for the treatment of waste sludges but are also suitable for stabilization of contaminated soils. Cement, lime, and other pozzolanic substances have been used as well as bitumen and resins.

In situ *treatments*

Treating contaminated material without removal has obvious attractions and there are a number of techniques for this. Simple cultivation may be sufficient but more complex approaches to the problem include leaching, grouting, isolation by barriers, vitrification, and electrokinetic techniques (Sanning 1985).

Cultivation Where a case of shallow surface contamination has been identified, it may be possible to plough the area deeply, buring the contaminated soil, by bringing uncontaminated subsoil to the surface. Patchy surface contamination might be dealt with by a mix-and-dilute policy in which contaminated soil is mixed with soil from clean areas to reduce the level of contamination below the trigger concentration. If contamination by acidity is superficial, ground limestone or basic steel slag may be spread and incorporated.

Leaching This involves the application of water or other solvents to the soil surface or, under pressure, below the surface. Contaminants are leached from the soil and the leachate collected for decontamination. This can be accomplished in suitably sized sections of a contaminated site. Leaching has been used to decontaminate a former herbicide factory site in Sweden (Lindfors 1980).

Grouting This is a familiar engineering process used to reduce the permeability of soil or jointed rock material below dam structures. Solidifying and gelling agents can be pressure-injected into the ground. The efficacy of the method depends upon the permeability of the soil. On derelict and contaminated land it has been used to control ground-water entry, to divert downslope drainage, to control gaseous emissions, and to limit or extinguish underground fires (Sanning 1985).

Barriers These can be inserted around contaminated areas by pile-driving, trenching, or grouting techniques to isolate toxic materials or control the movement of leachates (Fig. 6.2). By the use of a technique known as 'kerfing', a series of intersecting voids can be cut beneath an area of contamination and then filled with a suitable barrier material (Childs 1985).

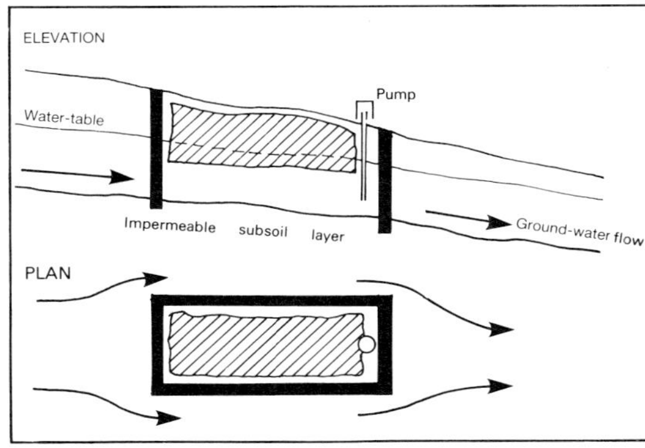

Fig. 6.2. Confinement of contaminant by barriers. (Reproduced by permission of Plenum Press and the author, from Smith 1985.)

Ground-water control Where an existing source of pollutants, such as a landfill, contaminates an aquifer, it may be possible to reduce or control the flow and volume of effluent by modifying ground-water levels. By the use of barriers and pumps the water-table can sometimes be lowered to avoid flow through the landfill and create conditions where attenuation of toxicity can take place beneath the source of pollutants (Childs 1985) (Fig. 6.3).

Vitrification It is theoretically possible to pass a massive electric current through the soil; the heat generated melts the soil and fuses metallic contaminants it contains. This technique is being developed in the United States for fusion and vitrification of contaminated ground. It is not yet a viable commercial method.

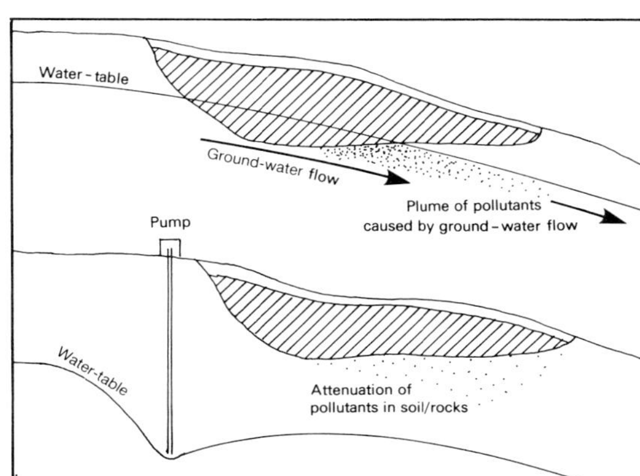

Fig. 6.3. Ground-water control of contamination. (Reproduced by permission of Plenum Press and the author, from Smith 1985.)

Electrokinetic techniques Electric current was used to desalinate soil in India as long ago as 1936 and the Russians have recently used similar techniques. A series of anodes and cathodes are inserted in the soil; water is added and the current applied. The saltwater is then collected at the cathode. It is thought that this method might be developed for decontamination purposes for certain soil pollutants.

Covering systems

The complete encapsulation of contaminated material is not often possible, so recourse is frequently made to covering systems (Parry and Bell 1985). An umbrella of impermeable material is placed over a contaminated site to control the entry of water, thus limiting the spread of contamination (Fig. 6.4). Various covering materials have been employed but well-graded, fine-textured soils are the most frequently used, except in climates where they are liable to become desiccated and crack. Soil materials may be modified by dispersants or mixed to broaden the particle-size distribution to reduce permeability. Synthetic covering materials range from rigid concrete slabs to flexible sheets or meshes of polypropylene. Sulphate attack can be a hazard for concrete; tearing, penetration, and exposure all affect plastic membranes. It is normal to protect plastic membranes with a layer of sand above and below to reduce rupture by sharp objects. Innocuous waste materials such as pulverized fuel ash can form a valuable covering material after leaching, or quarry waste, which comes in a wide range of particle size, also has been used. Dredgings from canals and docks are frequently polluted and should only be used after analysis has shown them to be uncontaminated; sewage sludge can be a useful amendment but may contain elevated levels of heavy metals, and builders' rubble can be useful especially when it is plaster-free.

In practice, most covering systems are usually made of several layers to serve different purposes (Fig. 6.5). A surface 10–15 cm soil layer capable of supporting vegetation is usually necessary, particularly on badly contaminated sites which are to be landscaped into grassed public open space. A subsoil layer allows deeper rooting and supplies moisture in dry periods. The barrier itself may comprise a layer of puddled soil or possibly a plastic sheet or mesh protected by sand. The barrier will normally have a domed or sloping surface to shed percolating water from the site and the upper layer of protective sand or gravel can act as a drainage zone. If venting is necessary, the sand or gravel below the barrier can provide an escape route for methane. Where there is a risk of toxic substances migrating upwards, a gravel or crushed limestone break layer may be necessary to interrupt the capillarity of the new profile. Crushed limestone is particularly effective as it neutralizes acids and precipitates heavy metals.

Recent cover systems described include: topsoil over gravel (as a break), over clay in Australia to cover zinc and copper waste in lagoons (Craze 1977); on a former refuse site near London, 150 mm coarse hardcore was laid for drainage and to inhibit the upward

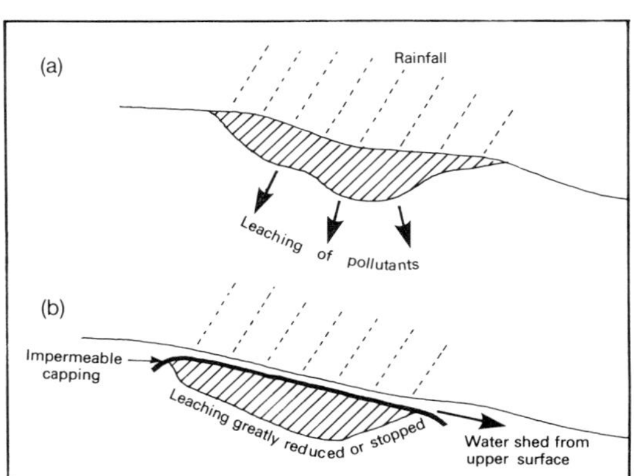

Fig. 6.4. Use of covering system to control leaching of pollutants.

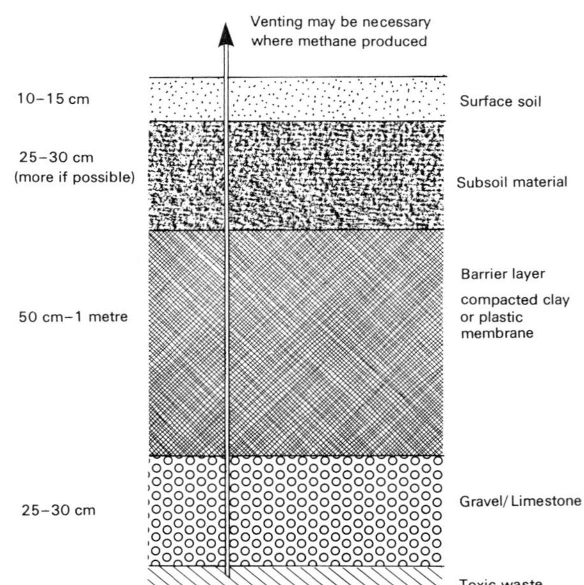

Fig. 6.5. Idealized covering system (not to scale).

movement of salts, and 25 mm of fine but porous material over this stopped material from the 150 mm of topsoil above from falling into the coarse hardcore; 550 mm of soil was provided where trees were to be planted, and a minimum of 550 mm of clean soil was stipulated for houses and gardens (McCarthy 1980); mine waste and tailings from lead and zinc mines in North Wales were covered with 100 mm of quarry rock waste plus a dressing of 5 tonnes per ha crushed limestone and the site seeded with a zinc–lead-tolerant clone of *Festuca rubra*, 'merlin' (Johnson *et al.* 1977) (Fig. 6.6); gasworks wastes at Beckton in London were covered with 225 mm of gravel to drain and to vent gases beneath a 1.2 m seal of London Clay covered by 300 mm of topsoil (Netherton and Tollin 1982); copper-refining wastes including copper, zinc, and cadmium in Cheshire were covered with 0.5 m of mixed calcium hydroxide–calcium sulphide wastes from the Leblanc process and then capped with 50 mm topsoil for a golf course (Jones *et al.* 1980); PVC membranes protected by a blanket of 100 mm sand and other layers have been used to cover domestic landfill in the USA and chemical wastes in Germany (Parry and Bell 1985).

Fig. 6.6. Restoration of former lead/zinc mine near Llanrwst, North Wales.

Types of restoration

Land may be restored for either hard or soft uses (Cowie 1981). 'Hard' use covers all sorts of building development whether industrial, public, commercial, or domestic. 'Soft' uses comprise amenity areas, public open spaces, and playing fields. Redevelopment for hard uses may require considerably more expenditure because buildings on disturbed and made ground require ground consolidation and piled foundations. There are also possibilities of the corrosion of foundations and services, and the presence of gas emanations.

Although the pattern of many site contaminants differs greatly from one site to another, restoration takes place for only a relatively restricted number of purposes. In rural areas restoration will be predominantly for agriculture, forestry, or amenity and the emphasis of the restoration must be to provide healthy growing conditions free from phytotoxic compounds. Urban sites are likely to contain a greater range of contamination problems and the restoration programme must mitigate the effects of toxic materials so that people can live and work safely. Healthy growing conditions are also required for vegetation where landscaping takes place. The design for restoration will be considered under the headings agriculture, forestry, public open space, housing, and industry.

Design for agricultural restoration

A plan to restore a derelict site for agriculture aims to provide the optimum growing conditions possible with the materials available. The new landscape should maximize gently sloping surfaces. The replaced topsoil should as far as possible be deep free-draining loam, sandy clay loam, sandy loam, or silt loam, with a high moisture-holding capacity and well supplied with plant nutrients or responsive to fertilizers. These conditions can probably be met in the restoration of sand and gravel workings in favoured lowland sites, but the soils available for replacement on most ironstone and coal reclamations are poor.

In England and Wales the restoration of agricultural land from opencast coal and ironstone workings has been strictly supervised since the Second World War. Indeed, as a result of the derelict land left by ironstone mining in the 1930s, the government appointed a committee (the Kennet Committee) which reported in 1939. It recommended that a restoration fund was set up and a mining operator had to deposit a bond based on the size of the site and to guarantee the payment of a levy on each ton of ironstone removed. The fund also received government payments and was sufficient to initiate reclamation. Thus dereliction was checked before it became widespread. When opencast coal mining began in 1942, similar regulations were instituted. Many sand and gravel sites worked before 1947 were not restored and little attempt was made to place restrictions on site operators to restore land after quarrying. This laxity has been replaced by more responsible attitudes, fostered by the Sand and Gravel Association which now has a good record of encouraging its members to do restorative work of a high standard.

Although opencast sites have never technically been derelict, they frequently occur where problems of derelict land exist. On the older parts of most coalfields, abandoned colliery sites, together with their underground derelict workings, often at relatively shallow depth, can be reclaimed and the coal remaining below ground extracted (Brent-Jones 1976, 1983). The restoration of land after opencast coal mining is therefore appropriate to this section.

Until 1958, compensation for the farmer's loss of his land was met under terms of the Compensation (Defence) Act 1939. This was replaced in 1956 by the Opencast Coal Act which gave the National Coal Board compulsory rights of temporary occupation and use of the land. These rights could be for ten years' duration, but the government ended the system in 1968 only to restore it again in the Coal Industry Act 1975, when plans were agreed for increasing the production from opencast mines. This Act of Parliament also required the operator to consult a wide range of interested parties such as the Ministry of Agriculture, the Forestry Commission, the Nature Conservancy, the National Trust, as well as the district valuer, for surface interests, and the minerals valuer for mineral interests. After due consultation has taken place the notice of intention to work must be placed in the *London Gazette* and in the local press.

The preliminary survey, undertaken before operations began, should have identified the broad relief and drainage of the site. Any proposed earth moving should be planned to achieve a natural form of landscape preserving existing trees and woodland. The location of services should be ascertained and diversions arranged where necessary, and underground shafts and foundations proven.

The second stage of assessment is a soil survey to record the character and distribution of the natural soils present. From this it is possible to plan the stripping of the topsoil which is removed and stacked. The subsoil is then stripped and stored as well as any other material which could be used to make up the quantity of subsoil necessary to re-cover the land to the required depth. The soil mounds are normally placed around the periphery of the site to screen the excavations from the surrounding property (Fig. 6.7) and no heavy equipment is allowed on them. They are

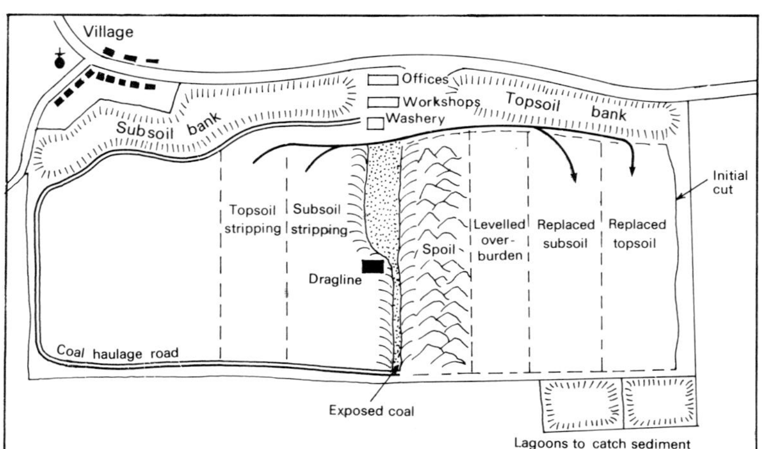

Fig. 6.7. Diagrammatic plan of an opencast site.

grassed and kept free of weed until required for replacement.

As the coal is removed, the draglines throw the overlying rock on to the area from where the coal was taken and bulldozers push it roughly into the landforms specified in the restoration plan. The shaping is continued by scrapers until the contours of the restored land merge satisfactorily with the land on either side. At this stage the contractor and an official of the Minstry of Agriculture consult on site to check and approve the relief.

As the equipment used for replacing the overburden is heavy, the whole area is subsoiled to at least 0.45 m, in two directions at right angles, to break up any material compacted by the passages of vehicles. Large stones revealed by this process are removed from the surface. Subsoil is then spread in two layers, each about 0.3 m deep and subsoiled to reduce compaction and to remove any large stones. Finally, the topsoil is replaced and cultivated to make it loose and aerated. Each stage is inspected by officials of the Ministry of Agriculture to ensure the work has been done correctly.

Once these operations have been completed and approved by the Ministry of Agriculture, five years of agricultural rehabilitation begins. First, the whole site should be laid down to grass to avoid erosion. Then the new layout of fields, hedges, and ditches must be decided. The relief of the site dictates the major divisions, but the details can be suited to the individual owner. Regulation of drainage water is of primary importance, and ditches are normally dug along the lines of the new hedgerows. Steep sections of the ditches are lined with stone to prevent scouring. Farmers tend to prefer quickthorn hedges round the boundaries of their properties where they are effective against trespassers, but wire fences are preferred for field boundaries within the farm as they require less maintenance. The farm roads, tracks, and footpaths are replaced and water is piped to stock watering troughs in the new fields.

The timing of operations is critical. Success or failure can depend upon suitable weather conditions when the soil is replaced. In the North of England it has been found best to replace the subsoil by the end of August and then to subsoil it thoroughly, and leave it throughout the winter to weather. Next spring, 7.5 tonnes/ha ground limestone is added and incorporated once the ground is sufficiently dry. The topsoil is then applied with a further 7.5 tonnes/ha ground limestone. It is cultivated in June/July, 630 kg/ha compound fertilizer added, and a seed bed prepared. A Cockle Park (University of Newcastle) grass mixture which includes perennial rye grass, cocksfoot, timothy, meadow fescue, and various clovers is sown. Once the seeds are established a top dressing of 250 kg/ha of nitrochalk is applied to maintain growth. A further 380 kg/ha compound fertilizer is added in the second, third, and fourth years with an additional 1.27 tonnes/ha basic slag in the third year (Brent-Jones 1976).

A permanent under-drainage system is installed in the late spring of the fifth year when the soil is dry enough to be worked without puddling. Tile drains are installed at 0.8 to 0.9 m at about 17-m intervals and covered with porous fill and then the land is mole ploughed at right angles to the drains. Finally, the land is cultivated and reseeded.

This system of management usually produces good crops and grass that can be cut and conserved as silage or hay. If grazing is allowed, it must be regulated to avoid any damage to the new sward through overstocking or through 'poaching' in wet weather. Stock should not be left on restored fields once the soil moisture content reaches field capacity in the autumn.

The cost of this supervised restoration is met by the National Coal Board. In 1975/6 it cost about £1235 per ha, as follows:

	Cost (%)
5 years agricultural management (cultivation, manuring, and cropping)	31.6
Permanent land drainage	27.5
Fencing	25.5
Ditching	2.5
Provision of water supplies	3.0
Planting hedges and woodlands	1.5
Farm roads and paths	2.4
Other expenditure	6.0
	100.0

In the East Midlands the stable, well-structured soils on the Northampton Sand Ironstone and other iron-rich strata facilitated restoration for agriculture. No heavy equipment was used and once the ironstone had been removed the soils could be replaced with little loss of fertility. Where the iron ore was overlain by limestones or boulder clays heavy machinery was required and restoration was more problematic (Bridges 1967).

Where the Oolitic Limestones and shallow soils

occurred above the iron ores, large blocks of limestone and insufficient soil remained for restoration. On a site near Colsterworth replaced soil was lost down cracks between the boulders. Elsewhere, the replaced material comprised boulder clays mixed with limestone blocks. On ironstone sites the rock materials are normally calcareous and free calcium may limit the availability of phosphorus and potassium which may reduce soil fertility (Fig. 6.8).

Blood *et al.* (1961) found that arable cropping on restored ironstone sites was unsuccessful without topsoil on two out of three experimental sites but that greater success was achieved with grass–clover or lucerne leys than other crops. Satisfactory growth was achieved with 150 kg/ha nitrogen, 70 kg/ha phosphorus and 125 kg/ha potassium. Where plant growth was possible, it was the practice to build up plant-available nutrients by putting land into ley pasture for ten years with high annual rates of fertilizer.

McRae (1981, 1983) describes the method of restoration employed on sand and gravel workings after extraction to a depth of 10 m. In 1977, topsoil and subsoil were first removed separately from a site and the gravels extracted. Subsequently the pit was filled with shallow layers of industrial waste, each compacted and covered with inert material and then finally sealed with 1 m of clay which was domed to facilitate drainage. The subsoil and topsoil were replaced in 1979 to give 1.2 m depth of soil over the domed clay cover. A grass ley including Italian rye grass was sown to protect the soil surface during the first winter and to encourage the formation of soil structure. In 1980, the site was cropped for hay with yields similar to those obtained from the site before disturbance.

A gravel quarry can be restored without importing landfill if there is sufficient depth of material between the floor of the pit and the water-table. In this case, extra material is excavated from the bottom of part of the quarry to raise the level of the land, leaving an area of open water for leisure activities. A drainage system is installed to control the water-table on the restored land. Unless the land is built up to 3 m or so above the water-table, such a site tends to be wet in autumn and winter, but a carefully timed programme of cultivation and sowing can produce yields as good as those before mining.

Design of restoration for forestry

In the British Isles, there have been few attempts outside Wales to restore land to forestry. Greater elevation, rainfall, and less fertile soils in the Principality make restoration for agriculture more difficult

Fig. 6.8. Restored ironstone quarry, Lincolnshire.

and limit the crops which might be grown. In such conditions, trees can be a useful crop on restored land. The problems encountered in earth moving are similar to those described for agriculture, so it is unnecessary to repeat the discussion of the previous section.

In the past, it has been the practice for engineers to produce a new landform with as low an angle of slope as practicable, often between 1 in 100 to 1 in 200 in order to minimize problems of erosion and gulleying. These low angles of slope safely discharge water from the site, but also result in saturated conditions in the rooting zone. This problem can be ameliorated by the construction of broad ridges, 30 m across and 1.5 m high in the centre (Fig. 6.9) (Binns and Fourt 1984).

Experience has shown that heavy earth-moving equipment compacts the ground and improved tree growth is obtained if the site is subsoiled to facilitate drainage and root penetration. A subsoiler has been developed by the National Institute of Agricultural Engineering for the Forestry Commission which has three heavy duty tines, the outer two having 'wings' which lift and loosen the compacted soil material. This process has to be done at optimum moisture conditions when the material is dry enough to be disrupted but not so dry that it breaks down to fine dust. In Britain the appropriate conditions normally occur during the summer months between May and September. Subsoiling takes place at right angles to the ridges at 75 cm depth. Where there is a porous substratum, water will pass off the ridges into the hollows and then into the rocks beneath. Where impervious conditions prevail, such as on a heavy clay material, subsoiling should take place to 50 cm depth and drains should be constructed along the hollows to take away excess water. Jobling (1981) demonstrated that survival and growth of trees is influenced mainly by the degree of looseness of the subsoil, the trees benefiting more from subsoiling in two directions. The use of valuable topsoil for restoring derelict sites to forestry is questionable, as amounts are limited and trees appear to root satisfactorily in replaced material without it.

An imbalance of available plant nutrients often limits tree growth on restored land. There is often sufficient potassium or magnesium but phosphorus, and particularly nitrogen, are deficient. Some nitrogen may be released from fossil reserves in shales during the first few years of weathering (Cornwell and Stone 1968; Aldag and Strzyszcz 1980), which ensures satisfactory establishment, but poor growth subsequently as it becomes exhausted. It is possible to fertilize tree crops in the same manner as for agricultural crops, but it is expensive, and if the topsoil is thin and the organic matter content low, the fertilizer is not retained within the rooting zone of the trees.

Experiments with trees growing on disturbed ground have shown that successive nitrogen dressings are essential to maintain good growth. Foresters have experimented with interplanting nitrogen-fixing trees and shrubs of which alder (*Alnus glutinosa*) and false acacia (*Robinia pseudoacacia*) seem to be the only tree species hardy enough. Unfortunately shrubs and herbs which fix nitrogen are palatable to rabbits, hares, deer, and sheep and are grazed out rapidly.

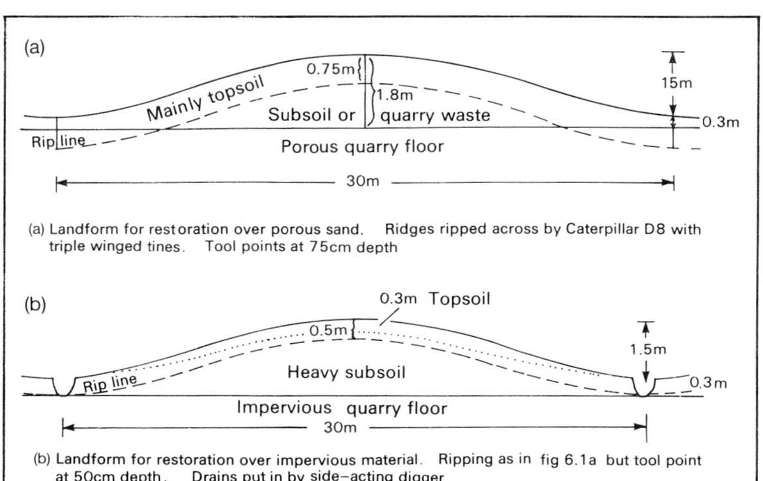

Fig. 6.9. Construction of landforms for restoration to forestry. (Crown Copyright, reproduced by permission of the Forestry Commission and the authors, from Binns and Fourt 1984.)

Plants like broom and gorse grow quickly but compete too effectively with young trees and are not recommended. Binns and Fourt (1984) recommend planting the tree lupin (*Lupinus arboreous*) and species of everlasting pea (*Lathyrus sylvestris* or *Lathyrus latifolius*) as these die out as the tree canopy closes and their accumulated nitrogen is then taken into the nutritional cycle of the tree crop. Sewage sludge has been used in many restoration schemes; however, at normal agricultural rates of 5 tonnes/ha/annum there are difficulties in spreading annual dressings without damaging the small growing trees. Davis (1982) therefore recommended 280 tonnes/ha in a single heavy dressing. There may be problems with the profuse weed growth which results. However, pockets of sewage sludge or moss peat have been used successfully to establish trees on very coarse quarry debris, such as the slate waste in North Wales. Once established, trees have a sufficiently deep rooting habit to obtain moisture from deep within dumps of coarse debris. Bradshaw and Chadwick (1980) reviewed the amenity planting of trees on waste materials in Norway, Wales, Australia, and Israel. They recommended local species, tolerant of poor conditions such as alder (*Alnus incana*), sallows (*Salix caprea* or *Salix aurita*), birch (*Betula pubescens*), Scots pine (*Pinus sylvestris*), oak (*Quercus petraea*) on acid rock wastes in Britain with ash (*Fraxinus excelsior*) and hawthorn (*Crataegeous monogyna*) on calcareous quarry wastes. Salt bush (*Atriplex* spp) or sage bush (*Salvia* spp) are recommended in drier climates.

Restoration to public open space

Often the most feasible solution to derelict land or badly contaminated land in urban areas is to convert it into public open space. Many derelict areas occur in urban areas where there are few parks or informal open spaces suitable for leisure activities. In the terms described earlier, public open space is a 'soft' use where slightly higher levels of residual contamination may be permitted. Most people's contact with public open space is limited; consequently, the risks associated with contamination are minimal. The use of public open space falls into two categories—informal recreation and formal activities requiring special pitches.

Informal recreation There are two components of the restoration of land for informal recreation which the surveyor must take into consideration when drawing up plans for informal public open space. The first is landscaping and the second is the plant cover necessary.

As has been described in previous sections, earth moving is achieved by large-scale machinery and most of the physical problems of reshaping can be solved with sufficient expenditure. If the preceding survey has revealed some particularly harmful materials on a derelict site, these may have to be buried out of reach of plant roots. It may be desirable or necessary to mix materials on a site to provide a more suitable growth media. Urban derelict sites may lack topsoil material and it may be necessary to use uncontaminated subsoil or to import soil (McCarthy 1980).

The landscape architect may use his imagination to produce a pleasing vista of hills and hollows with areas of shelter and seclusion. The designer should bear in mind that these areas will need maintenance afterwards and the drainage must be effectively integrated with the surrounding land. The major consideration is to produce a new landscape which has an aesthetically pleasing appearance (Hackett 1977). Normally, 0.3 to 0.5 m of cover material is considered satisfactory when the future land use will be for public open space.

After earth moving has taken place, it is normal practice to establish a grass cover as quickly as possible to reduce erosion. This may be followed by planting a mixture of trees and shrubs to provide a variety of vistas as the new landscape stabilizes. A proportion of nitrogen-fixing plants is desirable. The Forest Park at Stoke-on-Trent (Land Use Consultants 1970) is a good example, where a derelict landscape of marl holes, tips, and disused factory buildings has been converted into a pleasing area of informal open space.

The appearance of the restored landscape may be enhanced by the presence of a body of water if there are natural sources of clean water on site. In many cases mine subsidence may already have provided a 'flash'; such a feature can be used, enhanced by tree and shrub planting, to provide an additional range of habitats for wildlife and attract a wide variety of bird and animal life, which can be observed as part of the informal recreational activities. Unless the proposed site for a water body lies below the water-table, it will be necessary to line the floor of the basin with puddled clay to retain the water.

Formal recreation The pleasing and varied topography described for informal recreation is equally suitable for golf courses, by the addition of tees and greens. The possibility of settlement, contaminants, or gas emissions suggest that the club house should be located on undisturbed ground nearby and the same applies to changing rooms for other sports.

Pitches for team games need a smooth surface of about 1 ha, with just sufficient slope for drainage, usually between 1 in 70 and 1 in 80 lengthwise; 1 in 40 is tolerated across the pitch (Downing 1971). It is convenient to locate two pitches for winter games side by side on a platform of 2.0–2.5 ha which can then be used for cricket in summer. Surrounding spoil banks covered with soil and grassed provide viewing positions for spectators. The soils of the playing area will have to be imported, particularly for goal mouths and cricket squares where wear is excessive and play occurs in a range of weather conditions. An athletics ground with running tracks and a central area for field events may require over 1.8 ha and derelict land has been used for this purpose on the North Star site at Swindon and in the Lower Swansea Valley (Chapter 8, this volume).

Aquatic sites can be used for both formal and informal recreation. Gravel pits can be landscaped, flooded, and stocked with fish. The expanse of water can be used for sailing and, if sufficiently large, for power boats and water skiing; multiple use is possible if activities are kept separate in space or time.

Restoration for housing

The example of houses built over a chemical waste tip at Lekerkerke (p. 34) in the Netherlands illustrates the problems of building on contaminated sites. There may be similar difficulties on a range of former industrial sites. Unlike arable land, forestry, or public open space, human contact with garden and home is close and children are particularly at risk where contamination occurs. Digging, making mud pies, scuffing up dust, may all lead to the ingestion of potentially toxic material. Garden produce could be contaminated.

For housing, landscaping should maximize southerly aspects. If the site has sufficient relief, a range of levels may give a greater number of houses access to an attractive view. In landscaping the site, slopes of more than 1 in 20 should be avoided as they will require special work for adequate foundations to be laid, although feasible in an expensive layout. Without consolidation, building on disturbed land can be difficult as settlement causes cracking of walls which in extreme circumstances could make the buildings unsafe. It is advisable for houses built on disturbed land to have raft foundations and, if larger buildings are envisaged, these should be piled so that the foundations rest on undisturbed ground. Houses should not be built upon combustible materials.

Anyone who has moved into a new house will appreciate the problems of trying to create a garden from the rubbish left by the builders. On derelict land the difficulties are greater, often with contamination added to the problems. An example of disturbed land used for building occurs in the former ironstone field at Corby, Northamptonshire. Corby New Town Corporation, in conjunction with the Building Research Establishment, carried out investigations into the consolidation of formerly mined ground. Three experimental treatments were compared: inundation with water, surcharge with an added mass of rock waste, and dynamic consolidation by dropping weights on to the surface. The preliminary results of the experiment indicated that loading the disturbed ground with an added mass was the most effective (Rimmer 1976). Another area where former derelict land is being urbanized is at Telford, Salop (Whitcut 1981). Disturbed land can be used for housing developments but involves greater expense; normally, contaminated land should be avoided.

Potential housing sites where the soil analyses exceed the trigger concentrations require further examination and remedial treatment. There are three major options:

1. Excavate the contaminated ground and dispose of it elsewhere.
2. Deal with the problem of contamination on site or *in situ*.
3. Ignore it and do nothing.

Doing nothing is unacceptable to the general public and local authorities, except in the particular case where derelict buildings are visually attractive or archaeologically interesting and form a feature in a landscaped area, once made safe. Otherwise, there are but two alternatives.

If it is decided to excavate and remove the contaminated soil, the problem of where to put it arises immediately. Toxic wastes are permitted to be consigned only to 'special waste' dumps, and merely carrying toxic loads about can spread contamination through dust and accidental spillage. If possible, on-site remedial action should be considered first, as it may be possible to deal with a contamination problem simply by cultivation. The other possibilities include incorporation of hazardous materials into solid masses formed by cement, lime, lime/siliceous materials, organic polymers, or thermoplastic materials such as bitumen. These processes require the injection of setting agents into the ground but the results are often not uniform owing to extreme variability of materials on derelict land. Liquids may be injected into the ground to neutralize acids or alkalis or to inactivate other substances chemically. Water and other solvents may be used to leach out undesirable contaminants, but their application in clean-up operations has not been fully investigated.

On-site processing of contaminants typically involves excavation and some form of chemical or microbiological techniques to detoxify, neutralize, fix, or stabilize toxic materials. The most straightforward process is to incinerate, but this may be difficult in urban surroundings. So the most common means of dealing with these materials is by a covering system. In its simplest form, this operation involves capping contaminated areas with an impermeable barrier, the aim of which is to isolate the toxic materials. The addition of layers of gravel or crushed limestone is an attempt to interrupt the capillary rise of liquids, to neutralize acids, and to precipitate other substances. More complex situations may require lateral containment of wastes by barriers, and in some cases the ground below the wastes may require sealing. Although various forms of covering have been applied to hazardous materials, there have been few studies to assess their long-term efficiency and reliability (Stief 1985). It is for this reason that contaminated land normally should not be used for housing.

Reclamation for industry

Much restored derelict land in urban areas will be reused for industry. Modern entrepreneurs require large areas of level ground upon which factories, mainly for light industry, can be erected easily and provided with all essential services. Without these facilities, they will not be attracted to former areas of heavy industry, preferring 'green-field' sites instead. So, with government aid, 'advance' factories ranging in size from 0.8 to 2 ha are being built on restored sites by many local authorities. These factories retain the flexibility necessary for expansion should it be required.

As the risk of contamination upon human beings is minimal on such redeveloped sites, some latitude can be permitted in the levels of toxic materials present. In most cases, undesirable materials will be removed or buried and capped with a solid layer of concrete or asphalt. If the preceding survey was properly carried out, buildings can be located away from potentially dangerous areas, which can be encapsulated and left as a landscaped 'soft' area or made into car parks. Buildings of two or more storeys require foundations on piles that reach undisturbed ground, with precautions taken to prevent acid or sulphate corrosion. The generation of gas from landfill sites may possibly limit the development of industrial sites on many formerly derelict sites, although buildings are possible if adequate ventilation is provided (Carpenter 1986).

Any site where tipping has taken place will be costly to redevelop. Tipped materials tend to fall into the four categories: domestic waste; industrial waste; natural granular fill; and natural cohesive fill. These settle at different rates. Organic material decomposes rapidly with up to 20 per cent settlement, whereas natural granular materials compact by a small amount, usually less than 5 per cent. The rate and extent of settlement also depend on how the material was tipped, and the initial compaction. It may be helpful to know where particular materials were tipped and when each part of the area ceased to receive waste.

There are several methods of consolidation. Loading, saturation, and dynamic compaction were mentioned previously. Pre-loading is convenient as excess material can be piled in one place as a loading medium to compact the ground beneath. It should be left in place as long as possible. Lowering a high water-table can also increase the effective loading. Conversely, it is reported that loose fill may become consolidated by saturating it with water. However, mechanical compaction by heavy rollers, or traffic, is commonly used, but rolling is not suitable on soft clays or coarse demolition debris.

Other specialist techniques are employed where conditions warrant extra expenditure. Vibroflotation is a technique for shaking particles closer together. If gravel-sized material is introduced around the vibroflotation equipment, a dense 'stone column' can be induced through the unconsolidated material. Other vibratory methods employ compressed air, with the introduction of sand, to form a rigid column. Dynamic compaction can be achieved to depths of 15–20 m by dropping a heavy weight on to the surface from a height of up to 25 m.

Stabilization can also be achieved by injecting cement, sand, or pulverized fuel ash into unconsolidated material. This method has been used in Scotland and north-east England to stabilize mine dumps. Sand drains have been used since 1925 to dissipate pore-water pressure in saturated unconsolidated materials and to render them more stable.

Costs of reclamation

The figures of the Welsh Development Agency indicate the costs of reclamation where dereliction has been completely removed and the restoration is regarded as final. These figures cover a wide range of schemes and costs vary greatly according to difficulties encountered on specific sites. Table 6.1 gives the figures.

The average figure of £30 515 per ha for reclamation over the period 1978–85 covers gradually increasing costs. Before 1980 the average cost per hectare was in the region of £21 000, whereas the average for the years 1981–85 has increased to approximately £42 000 per ha. These average figures cover a considerable variation so the cost of some individual schemes is given in Table 6.2.

In general terms it would appear that in 1980 the costs of reclamation of uncontaminated land for agriculture and forestry were between £10 000 and £12 000 per ha, this has now risen to £18 000 to £20 000 per ha at 1985 prices. The costs of restoring contaminated land for recreational and amenity purposes are similar, but the higher standards required of a site to be redeveloped for industrial use or for housing involve expenditure of £35 000 to £50 000 per ha. Where contamination is severe and difficult to contain, restoration costs at 1980 prices may rise to between £80 000 and £100 000 per ha.

Table 6.1. Figures for reclamation of derelict land in Wales

Year	No. of schemes completed	Area (ha)	Cost (£)	Cost/ha (£)
1978–79	22	168	2.5 m.	14 881
1979–80	52	580	12.3 m.	21 207
1980–81	29	378	8.1 m.	21 428
1981–82	24	345	8.9 m.	25 797
1982–83	27	134	6.1 m.	45 522
1983–84	45	197	11.1 m.	56 346
1984–85	23	79	8.4 m.	41 774
Total		1881	57.4 m.	30 515

Source: Welsh Development Agency reports.

Table 6.2. Costs of individual schemes of reclamation

Year authorized	Scheme	Area (ha)	Cost (£)	Cost/ha (£)
1976	Parc Mine, Llanrwst	4.8	442 000	92 083
1977	Morfa Tinplate Works, Llanelli	1.6	82 000	51 250
1978	Tal-y-sarn slate waste	16.0	729 000	45 562
1978	Dyffryn Rhondda colliery tips	25.6	930 000	36 328
1978	ROF Pembrey	18.0	350 000	22 077
1979	Afan Valley Eagle Brickworks	26.0	574 000	19 444
1984	Cambrian Colliery, Clydach Vale	113.0	2 301 000	20 363

Source: Welsh Development reports.

7 The legal framework for reclamation of derelict land

Relevant legislation

Legislation about derelict land has been slow to evolve and particularly that on the provision of grant aid for land reclamation. In part, this has resulted from governments trying to link the clearance of derelict land with industrial redevelopment and job creation. Dereliction was not itself regarded as a significant problem until the second decade of this century. By the 1930s subsidence associated with coal-mining and dereliction left after ironstone mining led to the discussion of compensation for damaged land, but no thought of prevention or cure. The Special Areas (Development and Improvement) Act of 1934 contained some provisions on derelict land but failed to include some of the worst areas because of their relatively low unemployment rates. Consequently, the £2 m. earmarked by this Act had limited effect.

After the Second World War, the Distribution of Industry Act, 1945 removed these anomalies and for the first time money became available for reclamation of derelict land. The Board of Trade was empowered to acquire derelict land lying within 'development areas' and to carry out any work necessary to return the land to usefulness or to improve the amenities of the neighbourhood. The good intentions of this legislation were defeated by national financial strictures and by the requirement in many derelict areas that restoration should alleviate unemployment, which still remained a condition for receipt of grants. In most areas, it could not be demonstrated that reclamation would lead to lower unemployment. It also proved difficult to obtain grants because the definition of derelict land required that the land be of no value to the owner and few were willing to admit their land was worthless or to hand it over to the local authority for reclamation and redevelopment.

The National Parks and Access to the Countryside Act of 1949 allowed grants of 75 per cent of the cost of reclamation necessary for the improvement of amenities, but only in National Parks or Areas of Outstanding Natural Beauty.

The linkage of reclamation and increased employment continued in the Local Employment Act of 1960 which combined the powers contained in the 1945 Act with a wider definition of derelict land, which now included neglected or unsightly land for the first time. The Ministry of Housing and Local Government was empowered to make grants within local employment exchange areas, thus changing the emphasis from regional to assistance of a more localized nature. However, the chronic economic problems of north-east England and central Scotland eventually forced a return to regional policies in the 1966 Industrial Development Act. This again recognized the development areas of the 1945 Act and offered grants of up to 85 per cent of the costs of reclamation of derelict land. Parallel legislation, the Local Government Act of 1966, spread the availability of financial aid to the reclamation of derelict areas outside the development areas but only up to 50 per cent of the cost would be given. Both Acts of 1966 offered financial assistance to local authorities to acquire derelict land by agreed purchase, or by compulsory purchase if necessary. The return to the policy of development areas made grants available for lands formerly excluded, but it still restricted grants to areas where reclamation would result in increased employment opportunities.

1966 marked a turning point in attitudes to derelict and despoiled land for in that year 144 people, mostly children, were killed when a coal-shale tip became unstable and engulfed part of a school in the village of Aberfan. The sympathy of the public was sharply focused upon the victims and cause of the tragedy. With strong public backing, the government acted quickly and within 24 days the Secretary of State for Wales announced to the House of Commons that a special unit had been created to deal with the problem of derelict land in Wales.

Under the provisions of the Local Employment Act of 1970, certain areas of Britain were given the status of 'special development areas', 'development areas', 'intermediate areas', and 'derelict land clearance areas', within all of which grants of up to 100 per cent of the cost of reclamation were made available. National Parks and Areas of Outstanding Natural Beauty qualified for 75 per cent grants and the remainder of the country qualified for 50 per cent grants (Fig. 7.1).

A Local Employment Act and an Industry Act, both of 1972, superseded the 1966 Acts already mentioned. As a result, it was no longer necessary to obtain a certificate from the authorities which stated that restoration would enhance employment prospects by enabling the creation of new industry. The Secretary of State for Trade and Industry was able to designate any area to be a derelict land clearance area, even though it was outside the assisted areas depicted in Fig. 7.1. Grants were increased to 85 per cent of the cost of acquiring the land and reclaiming it. Thus the traditional link between the creation of employment and the rehabilitation of derelict land was broken and derelict areas could now be restored solely for amenity or environmental reasons. Grants became available for 20–30 per cent of the costs of roads and sewers on restored sites.

In 1975, Development Agencies were created in Scotland and Wales to provide sites, factories, and investment capital for new or expanding industries. These agencies were given powers to set up business advisory services and to acquire derelict land to improve the character of the environment. The legislation which set up the Welsh and Scottish Development Agencies repealed the provisions of the 1972 Local Employment Act as it applied to Wales and Scotland, but the still applied in England. In 1976, the Welsh Development Agency made grants of 100 per cent for reclamation; this could include the cost of acquiring the site, necessary surveys, restoration, but was nett of the value of the site after restoration.

Since the United Kingdom joined the EEC, funds

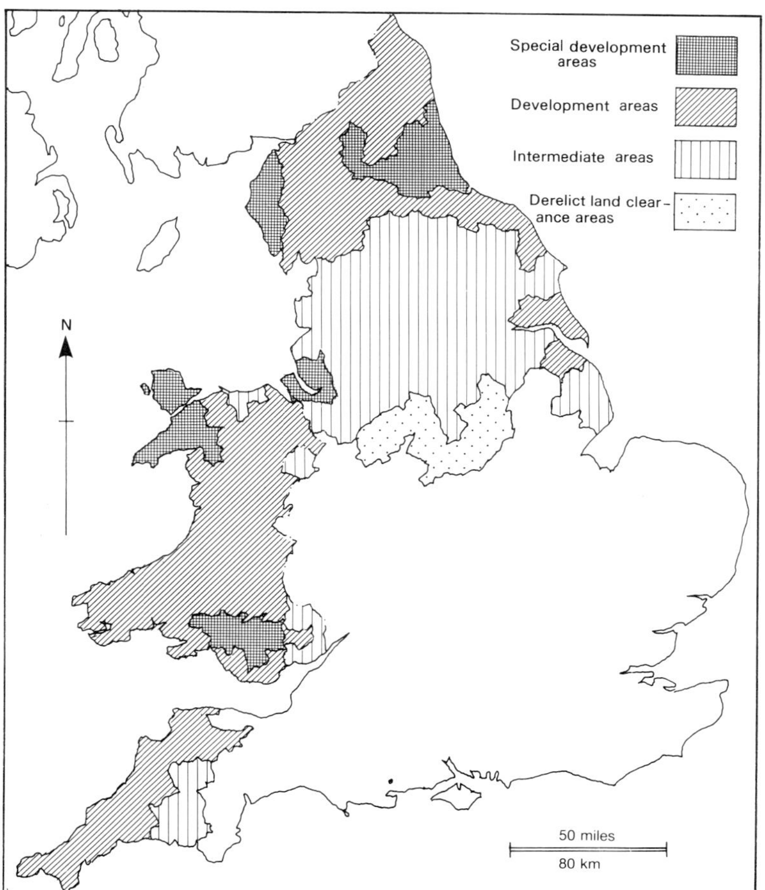

Fig. 7.1. Areas of England and Wales with designated status for development in 1977. [Redefinition of these areas has occurred since, the most recent in 1982. (Crown Copyright.)]

have become available from the European Regional Development Fund (ERDF) established in 1975 to help correct economic imbalances in the countries of the Community that resulted from economic change. The ERDF provides assistance for regional projects in industrial (including tourism), handicraft, or service activities which are in receipt of government regional aid. It also assists infrastructure works related to regional development where these are carried out by local or other public bodies. The assisted areas mentioned previously and shown in Fig. 7.1 were eligible to take advantage of this fund. Similarly defined areas in other EEC countries also qualify. In particular, the fund can assist the development of regional infrastructure giving access to industrial areas and the realignment of electricity, gas, telecommunications, and railways also qualify. Schemes with a high industrial or tourism element are most favoured.

It is the intention of the ERDF to assist new projects by meeting the costs of site investigation, site clearance, construction work, plant and machinery, consultants' fees for design and supervision, landscaping, other capital costs (wayleave, disturbance, and severance payments), and certain land costs. The emphasis is on larger schemes for public benefit rather than small ones where individuals stand to gain. Theoretically, clearance of derelict land for industrial purposes is eligible, but no further benefit can be obtained where 100 per cent of costs are met by grant aid. However, where local authorities reclaim land under the Local Employment Act, 1972, the ERDF will aid infrastructure developments once the land is cleared. Many local authorities have benefited from this combination of financial aid.

Certain environmental schemes in industrial improvement areas in designated districts may receive 75 per cent grants under a Local Government (Social Need) Act of 1969 and the Inner Urban Areas Act of 1978, as summarized in a DoE Environment Circular (DoE 1977).

The Local Government, Planning, and Land Act of 1980 made 50 per cent capital grants for reclamation available to an individual. Previously, grants had been restricted to local authorities and land owned by them. This change was designed to stimulate involvement by the private sector in reclamation, especially in urban areas. The grant was payable to site owners on completion of reclamation having regard to the increased value of the land. This Act also introduced a public register of derelict land with the aim of accelerating the release of land for redevelopment.

Since 1981, all local authority reclamation works have been placed in one of two categories: **Category A** schemes are designed to involve the private sector in developing industry, commercial premises, or housing immediately after reclamation. The developer must make a formal declaration to buy the land after reclamation. Where land reclaimed under a Category A scheme is sold to a developer, the local authority is allowed to retain part of the grant aid to fund further reclamation projects. **Category B** schemes are those undertaken by local authorities for the environmental improvement of large areas, such as the sites of former steel works.

A Derelict Land Act came into force in 1982 to consolidate statutory provisions for dealing with derelict land. This Act increased the level of grants to 80 per cent for private individuals and 100 per cent for local authorities.

In addition to the legislation directly relevant to derelict land, there are many other Reports and Acts of Parliament which must be taken into account. Thus, the Kennet Committee (1939) looked into dereliction resulting from ironstone mining. Its findings were directly responsible for the establishment of the Ironstone Restoration Fund in 1951, under which the costs of reclamation were borne partly by the mineral extractor and partly by the government.

This precedent undoubtedly helped to restrain the potential dereliction when opencast coal mining was introduced in 1942 to boost coal production. A Technical Advisory Committee appointed by the Ministry of Agriculture drew up a reclamation code which came into force in 1951. This specified the operating conditions for stripping, storage, and replacement of topsoil and a five-year period of after-care for sites.

In 1972, the Stevens Committee (Stevens 1976: date of publication) reported on planning control over mineral workings and their findings eventually passed into law as the Town and Country Planning (Minerals) Act of 1981. Few, if any, conditions for restoration were placed on permissions to extract gravel in earlier years, but now up to 50 detailed planning conditions may be appended, many of which relate to site restoration.

The 1981 Act tightened many of the intentions of the Town and Country Planning Act of 1971; local authorities are now required to keep mineral extraction sites under review. In addition, a local authority can insist on adequate after-care arrangements and suspension and prohibition orders may be issued during the life of a quarry if the regulations are contravened. All permissions are for a specified period of operations, not exceeding 60 years.

Other statutory controls concern the waste materials which often occur on derelict sites. The Control of Pollution Act of 1974 refers to most industrial and commercial wastes but excludes the waste from mines and quarries which is covered by the Mines and Quarries (Tips) Act of 1969. Disposal of waste may take place only where planning permission has been given under the Town and Country Planning Act of 1971 in England and Wales, 1972 in Scotland. The disposal of poisonous substances has to be notified to the appropriate authority under the Disposal of Poisonous Waste Act of 1972. The 1974 Control of Pollution Act also includes provisions for dealing with radioactive substances, formerly dealt with under the Radioactive Substances Act of 1960. Movement of any radioactive substance requires official permission. Movement of materials with a low level of contamination must be dealt with under the Ionizing Radiation (Unsealed Radioactive Substances) Regulations of 1968, now superseded by Ionizing Radiation Regulations, 1985.

Finally, the Health and Safety at Work Act of 1974 has considerable implications for the reclamation of derelict land, and requires appropriate action to be taken to protect the work-force and also persons who are not in employment but may be affected by work activity. Employers should be aware of its implications.

Applications for financial assistance for derelict land clearance

Up to 1980, the acquisition of derelict land by local authorities was a prerequisite to any restoration programme, as they alone could be the recipients of grant aid. Applications must still be chanelled through the local authority even where a scheme is being undertaken in partnership with private individuals, possible since 1981. The local authority must apply to the Department of the Environment in England, and the Welsh or Scottish Development Agencies in those countries. The following comments are based on advice given to local authorities by the Welsh Development Agency but the principles remain the same for England and Scotland.

Assuming that a derelict land survey has taken place as described in Chapter 5, the information should be available for the local authority to prepare an application for funding. The first stage is to establish the acceptability of the proposed reclamation scheme. This preliminary submission should include:

1. A description of the land indicating why it is considered derelict, neglected, or unsightly.
2. Details of any adjacent land which might be required temporarily during the period of reclamation.
3. Whether the land is held by the local authority or is about to be acquired so the work may be carried out; the powers which are to be used for acquisition should be clearly stated.
4. A site location plan at a scale of 1:10 000.
5. A broad outline of the work to be carried out should be given together with estimates of the cost of works, including acquisition, earth-moving, drainage, landscaping, and the cost of design and administration.
6. The proposed after-use.
7. The presence of any features of archaeological interest.

Once the preliminary submission has been accepted by the DoE, Welsh (WDA), or Scottish (SDA) Offices, the second stage of application may follow. This comprises a detailed submission and includes design work for reclamation in sufficient detail for a formal application for funding. This formal application should include:

1. Any changes in the area which have taken place since the preliminary application.
2. The existing and proposed landforms should be indicated with plans to show the proposed drainage and the proposed after use of the site. Plans and cross-sections may be required to illustrate certain points in the design of the reclamation.
3. Design calculations for the reclamation may be required by the DoE, WDA, or SDA.
4. An estimate of the total cost of the proposed scheme, including the site engineering works, supervision, design fees, land costs, legal, and other administrative fees. The estimated cost of engineering works should be based on a draft bill of quantities which should be submitted.
5. It should be confirmed that all necessary planning consents have been obtained or are unnecessary.
6. If the land has been acquired, the valuer's reports should be forwarded with the formal application, but if the purchase has not been concluded, an estimate of the valuation may be used on the understanding that if the scheme is approved, grants will be made on the basis of actual costs.

Only when the local authority is informed that the scheme is accepted may it proceed to draw up the documents for tender. After advertising or circulation to a list of acceptable contractors, the tenders are considered and the successful tender forwarded to the grant-aid body. If the tender is found to be acceptable to the grant-awarding agency, approval is given to the local authority, who in turn, may accept tender and issue an order to commence work.

Monthly reports on work in progress are required and any increases in costs should be mentioned in these reports. Claims for payment should also be made at monthly intervals until the final cost is known. A broadly similar procedure is followed in applications for funds from the EEC's Regional Development Fund (ERDF), with the DoE, WDA, or SDA acting as intermediaries for passing applications to Brussels.

8 Case histories

There is a growing body of expertise on surveying and restoring derelict or contaminated land. Unfortunately, details of this expertise are not widely available and occur mainly in conference proceedings or scientific reports of limited circulation. For this reason case histories are presented of four reclamations. The Lower Swansea Valley Project was the first multidisciplinary approach to such a problem in Britain. Restorative work in the valley is now approaching its conclusion almost 25 years after the inception of the scheme. The Rising Sun Scheme at Blaina covers a mixed coal-mining and industrial site which was the largest single coalfield restoration in Wales. It can be paralleled by other schemes such as the reclamation of a site occupied by several collieries at Shipley in Derbyshire which has been turned into a country park. The third example illustrates the problems encountered upon former railway land involving the difficulties of underground combustion and metal contamination. Finally, the problems of the London Docklands and Thamesmead bring together several of the difficulties experienced on a large site with a multipurpose history.

The Lower Swansea Valley

The 480 ha of industrial dereliction in the Lower Swansea Valley resulted from almost 250 years of smelting and processing metals. At different times, copper, lead, silver, arsenic, and zinc were smelted in 22 works along the tidal reach of the River Tawe. Later, 10 steel and tinplate works were established further inland (Fig. 8.1). Up to 1880, non-ferrous metals were most important with two-thirds of the copper imported into Britain smelted in this one area. Copper smelting declined after 1880 and was replaced by zinc. After 1928, steel and tinplate dominated the industry of the valley and four out of every five British tinplate workers were employed within 20 miles of Swansea. The legacy of dereliction included virtually all the characteristic features such as seven million tons of slag, ruined buildings, restricted access, and unvegetated, contaminated, eroded soils on the valley side (Figs 8.2 and 8.3).

After the Second World War a predominantly industrial future for the valley had been envisaged when major industries were still active. The local authority had prepared plans for redevelopment under the Town and Country Planning Act, but lack of finance for restoration prohibited any action. The Lower Swansea Valley Project was set up in 1961 with the financial assistance of the Nuffield Trust, Swansea County Borough Council, the Welsh Office, and the University College of Swansea. The brief of

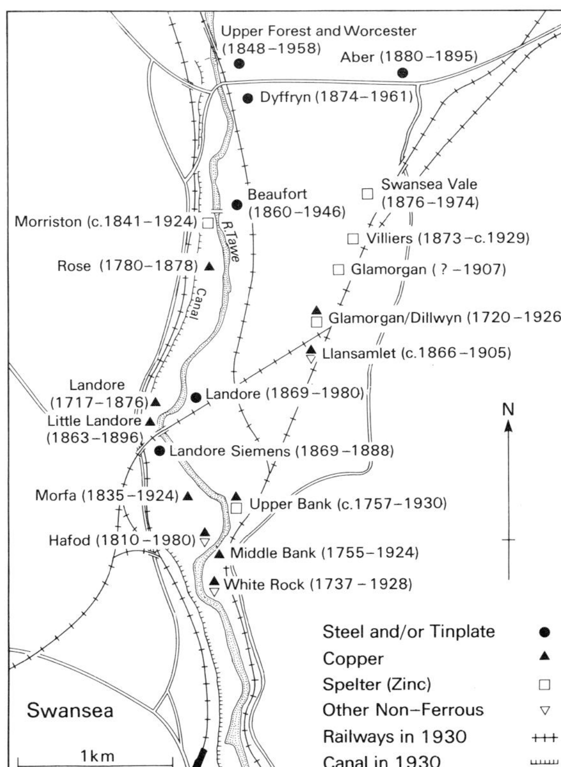

Fig. 8.1. Former metalliferous works in the Lower Swansea Valley.

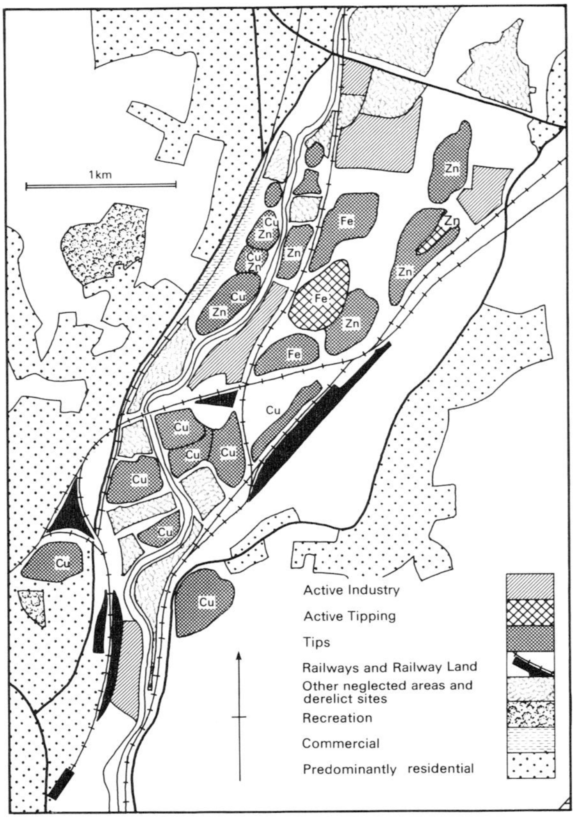

Fig. 8.2. Waste materials and land use in the Lower Swansea Valley, 1962.

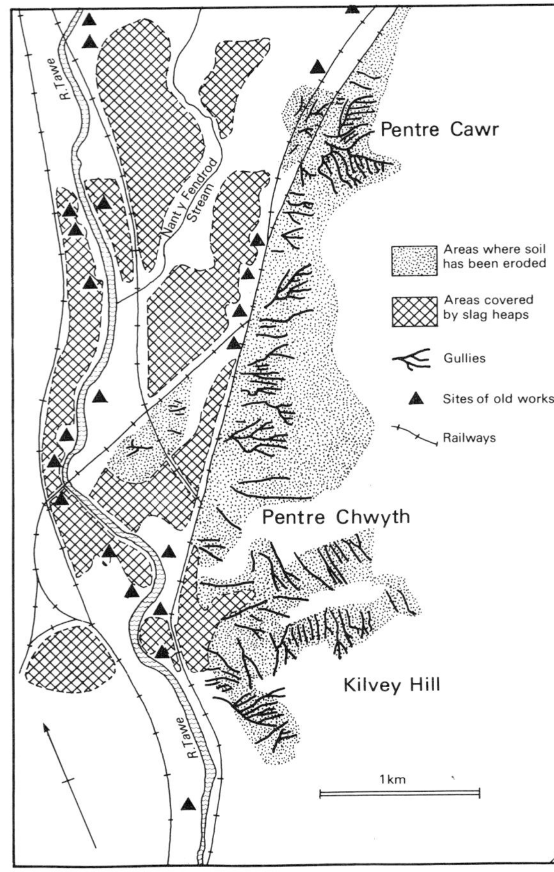

Fig. 8.3. Soil and gulley erosion in the Lower Swansea Valley, 1965.

the project was '. . . to investigate the physical, social and economic situation in the Lower Swansea Valley, to understand the reasons which had inhibited its development in the past and to provide the information necessary for its future development'. The project was clearly seen as a first stage in which information was to be gathered and interpreted, leading eventually to the renewal of the devastated land and development of new forms of land use. The work of the Lower Swansea Valley Project was described by Hilton (1967) and Barr (1969).

The findings of the project (Fig. 8.4) were incorporated in a draft development plan of 1968 which also included improvements to the river banks. This plan in due course was subsumed into a consultants' report published in 1974 which recommended a phased development of a mixture of industrial and leisure activities accompanied by landscape improvement. After local government reorganization in 1974, the new City Council published an Interim Planning Statement (1975) for the valley which included a 'Forest Park' and a 'Riverside Park' which reflected a more positive attitude to the provision of facilities for leisure activities. These ideas grew into the 'Five Park Scheme' (1981) in which the Enterprise Zone, designated by the government in 1980, became the Enterprise Park for industrial activity, accompanied by a Leisure Park, a Riverside Park, a City Park, and in the old dockland area, a Maritime Park (Fig. 8.5). Thus the restoration of the Lower Swansea Valley has taken place within a wider framework of urban renaissance than originally anticipated by the pioneers of the Lower Swansea Valley Project.

Investigation of the physical environment

Existing Ordnance Survey maps showed insufficient detail to provide a basis for the work of the project. Therefore, aerial photography was obtained in June

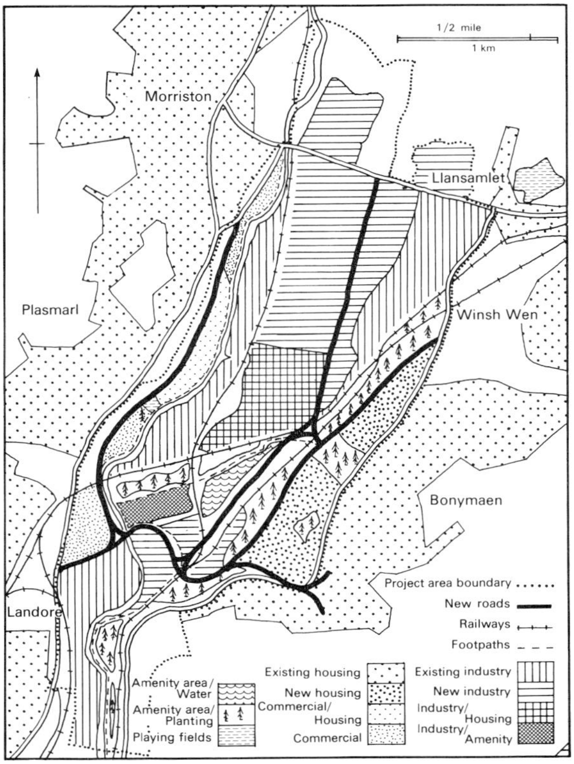

Fig. 8.4. Land use proposals for the Lower Swansea Valley, 1967.

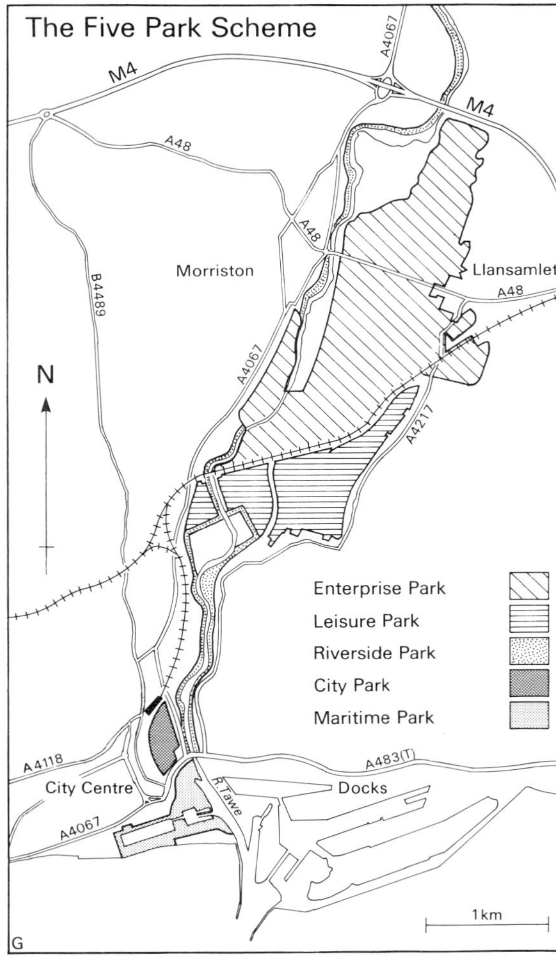

Fig. 8.5. Swansea City Council's Five Park Redevelopment Scheme.

1962 and a detailed map at 1:5000 compiled by photogrammetry with a contour interval of 10 ft up to 200 ft OD and a 20 ft interval above 200 ft. The positions of former smelting works on the existing 1:1250 Ordnance Survey plans were superimposed on to the photogrammetric map. The detailed base map allowed calculations of the area and volume of the many tip complexes. To assist this work and the subsequent geological, ecological, pedological, and hydrological contributions, the project area was subdivided into plots where specific problems occurred. These plots subsequently formed the basis of the reclamation plans.

In the early 1960s it was thought too expensive to remove the slag heaps, so experiments were begun to find how plants might be encouraged to grow on metalliferous wastes. If the tips could be covered with plants, this would be a relatively inexpensive way of improving the visual appearance of the valley (Street and Goodman 1967). Amendments such as sewage sludge, domestic refuse, pulverized fuel ash (pfa), and inorganic fertilizers were investigated in a series of field and plot experiments with mustard, common bent grass, and a grass ley mixture. This work was essentially empirical as little was known of the physicochemical factors which controlled the availability and uptake of heavy metals by plants. Growth was best on the steel wastes and poorest on zinc waste, with copper intermediate (Weston *et al.* 1965).

Trials were made of different species of shrub known to be tolerant of acid conditions, atmospheric pollution, drought, or exposure in experimental areas on copper, steel, and zinc waste tips. They include *Ligustrum vulgare, Buddleja davidii, Hippophae rhamnoides, Rhododendron ponticum, Salix repens, Clematis vitalba, Lupinus arboreus, Robinia pseudo-acacia, Sorbus aucuparia, Thelycrania sanguinea,*

Alnus glutinosa, Betula verrucosa, Medicago (innoculated with *Rhizobium* sp.), and *Tagetes minuta*. Some of these plants are natural colonizers of derelict areas, and some are nitrogen fixers.

The plant species growing naturally upon the various tip materials in the Lower Swansea Valley were surveyed (Table 8.1).

There was a sparse cover of common bent grass (*Agrostis tenuis*) and wavy hair grass (*Deschampsia flexuosa*) upon the stony subsoil material exposed in the eroded soils surrounding the smelters (Bridges 1969) (Fig. 8.3).

In parallel with this work, plant species which had evolved a natural tolerance to heavy metals were collected from soils known to be rich in copper, lead, or zinc. By cultivation and seed collection it was hoped to obtain sufficient seed from tolerant plant clones to use in revegetation. This work showed promise but it was not possible to complete it during the Project. Subsequently the National Seed Development Organization produced 'merlin', a variety of *Festuca rubra* which was lead–zinc tolerant.

The microbiology of the three main tip types in the Lower Swansea Valley was investigated. Bacteria of several groups were isolated, but not the nitrogen-fixing *Azotobacter* not mycorrhizal fungi. Measurements of soil respiration suggested that soil-living micro-organisms were inhibited more through lack of organic matter than by metal toxicity.

Originally, a major aim of the Project was to improve the appearance of the environment by planting trees on the eroded natural soils. From 1963 onwards, Japanese larch, Lodgepole pine, and birch were planted on 37 ha after liming and fertilizing. The trees were provided by the City Council and the Welsh Office and planted at minimal cost by volunteer labour. These trees now make a major contribution to the improved appearance of the valley (Fig. 8.6). (Certain areas were planted with grasses but with less success.) Since 1973, the emphasis has been on landscaping in the Enterprise Zone, paid for by the Welsh Development Agency from government funds and aided by grants from the European Regional Development Fund as part of infrastructure development.

Social aspects

Investigations of the human environment were a significant part of the initial research. A transportation study found that the valley floor was almost inaccessible by road and large areas were only accessible on foot. The River Tawe formed a barrier along the western side of the project area and railways and disused canals broke the land into several small, awkwardly shaped areas (Fig. 8.2). Recommendations for a new road network for the valley included a spine road to serve the northern part of the project area, a new crossing of the River Tawe in the Hafod–Pentre Chwyth region, and improvements to the roads on both sides of the valley. The western road improvements were to follow the line of the Swansea Canal to bypass Morriston.

A social survey of the valley and surrounding districts examined the distribution of sexes, marital status, occupational status, educational attainment, income, children in receipt of free school meals, children taken into care, and juvenile delinquency in the population. The housing stock was found to be old and in need of replacement or renewal. Schools and other public buildings were inferior to those elsewhere in the town. There was a shortage of public open space and it was recommended that this should be taken into consideration in the preparation of the development plan.

The 'park' concept is one of the most significant features of final development plans drawn up by the Swansea City Planning Department which goes a long way to meeting the earlier proposals for public open space in the valley. One such park is the Leisure Park

Table 8.1. Plant species occurring on tip materials in the Lower Swansea Valley (Weston *et al.* 1965)

Tip material	Ground covered		Total number of species recorded	Number of quadrats	Average number per quadrat
	Including mosses (per cent)	Excluding mosses (per cent)			
Zinc waste	10	1	4	50	0.04
Copper waste	27	4	15	57	1.5
Steel waste	16	3	36	70	3.2

Fig. 8.6. Afforestation in the Lower Swansea Valley, 1978. (Photograph includes much of the devastated area shown in Fig. 1.5.)

occupying land south of the main railway line with facilities for informal activity and organized sport.

Restoration

Although small areas had been restored, such as the Llansamlet arsenic works site alongside the main railway line, no significant reclamation took place before 1966 when government financial assistance became more readily available. Reclamation has continued since then, culminating in the last major scheme which began in 1982. The reorganization of local government in 1974 gave fresh impetus to the process of restoration as the new Swansea City Council gave priority to the work. A small executive committee was set up with power to take decisions without going through the normal lengthy channels. As a result, full advantage has been taken of any financial aid, with the City Council able to respond rapidly and positively to any opportunities for redevelopment.

The 1967 Lower Swansea Valley Report recommended acquisition of land as a fundamental requirement for redevelopment in the valley. The local authority adopted this policy and proceeded to obtain the freehold of land by negotiation. All the land recommended by the Report has now been acquired by the City Council. This policy paid dividends as the Welsh Development Agency would only accept applications for improvement grants for land which was in public ownership. (Only after the 1982 Derelict Land Act could private individuals qualify for assistance with the clearance of derelict land.)

Reclamation began on sites on the periphery of the valley, where there was access for heavy machinery, and gradually worked towards the centre of the derelict area. The sequence of reclamation schemes is given in Table 8.2 and Fig. 8.7, the work being done under contract by outside contractors but administered and supervised by the City Council Engineering Department. Schemes have been initiated only when financial assistance has been assured and the progress reflects the provision of government funds (Bromley and Morgan 1983).

In 1967, it was estimated that clearance of derelict land from the Lower Swansea Valley would cost £2.258 m. or £3077 per ha. Between 1967 and 1985, £4.8 m. was spent on restoration schemes, most of which was received as grants from government sources. Major schemes of derelict land clearance in the Lower Swansea Valley began in 1967 with the removal of 183 000 m^3 of copper slag from the 33 ha White Rock tip. This and the landscaping of the site for amenity purposes cost £129 210 or £3915 per ha. The slag was used to raise the level of the land at the Upper Forest and Worcester steel and tinplate works

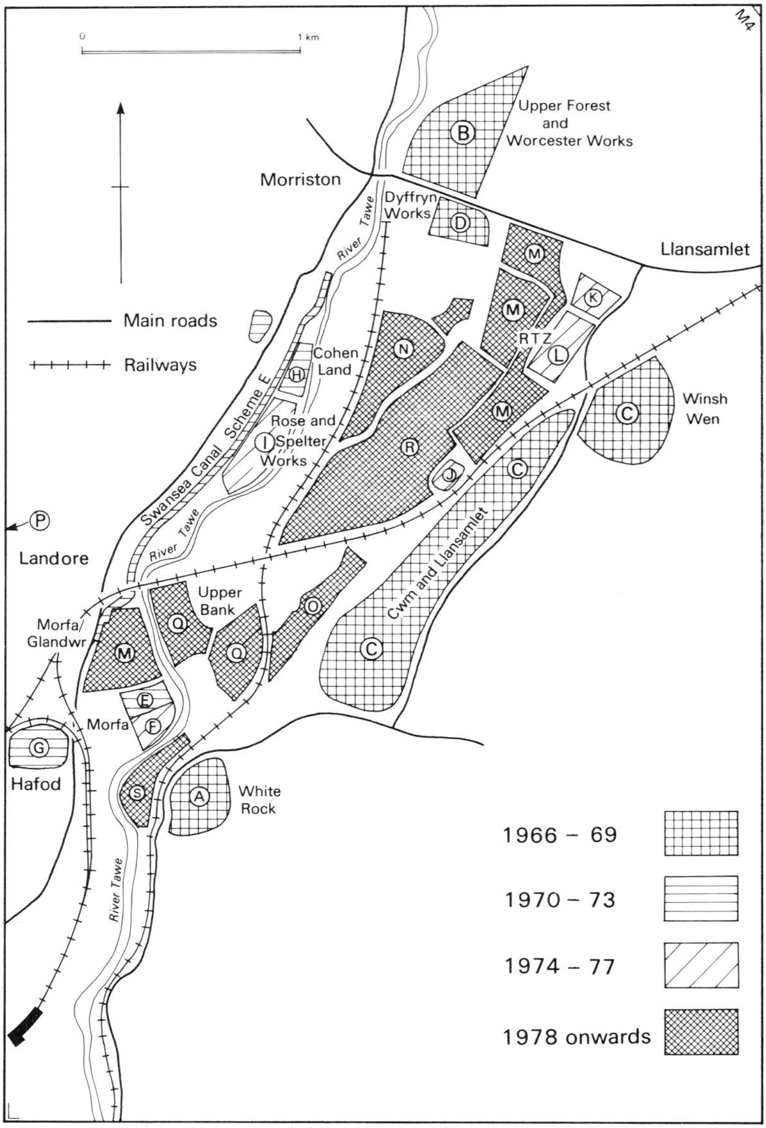

Fig. 8.7. Reclamation schemes completed between 1967 and 1985.

on the valley floor to reclaim 16 ha for an industrial site at a cost of £295 391 or £18 461 per ha. The adjacent 8 ha site of the Dyffryn steelworks was cleared in 1969 for £131 507 or £16 438 per ha, also for industrial purposes. However, the removal of the Pentre–Hafod tips of copper waste to clear a small area for school buildings cost £403 616 or approximately £80 723 per ha (Bromley and Humphrys 1979).

In 1979, a conference was held to review progress and to encourage the completion of the task of rehabilitation of the valley (Bromley and Humphrys 1979). Although weighted strongly in favour of planning and social studies, this conference indicated that research into the environmental background was continuing but at a lower intensity (Bridges *et al.* 1979). Swansea City Council had commissioned studies on the River Tawe, the extension of Pluck Lake as an amenity feature, and on basic development plans. Further aerial surveys were flown in

Table 8.2. Reclamation schemes in the Lower Swansea Valley (1967–85)

Year	Map code	Name of scheme	Area (ha)	Volume (m^3)	Cost (£)
1967–68	A	White Rock tip	33	183 000	129 210
1967–68	B	Upper Forest/Worcester works	16	200 000	295 391
1969–70	C	Cwm, Winsh Wen and Llansmalet	104	65 000	109 289
1969–70	D	Dyffryn steelworks	8	50 000	154 732
1970–71	E	Swansea canal/Morfa I	11	180 000	162 567
1970–76	F	Morfa II	4	63 000	110 204
1972–73	G	Pentre–Hafod tip	5	112 000	403 616
1973–74	H	Graig brickworks/Cohen	3	42 000	32 562
1974–75	I	Rose and Spelter works	16	350 000	105 455
1975–76	J	Glamorgan works	0.5	—	4 845
1975–76	K	RTZ phase I	3.5	75 000	58 000
1976–77	L	RTZ phase II	5.5	80 000	248 888
1978–79	M	RTZ phases III and IV and Morfa/Glandwr	24.5	620 000	930 466
1978–79	N	Lower gas works	13	189 000	166 350
1978–79	O	Pluck Lake tip	15	32 000	30 000
1978–79	P	Pentre tip	3	86 000	48 000
1981–82	Q	Upper Bank tip I and II	15	1 172 000	329 712
1982–86	R	BSC tip and lake	46.7	200 000	1 434 000
1984–85	S	White Rock works	10	75 000	107 000
Total			336.7	3 774 000	4 860 287

Sources: Elias 1979; Meller 1979; Jones 1979; Bromley and Morgan 1983; WDA Annual Reports.

1969 and 1982. With the announcement of the Enterprise Zone in 1980, investigations of bearing capacity of certain sites for industrial development have been undertaken specifically with the aim of providing information for prospective site occupiers.

In retrospect, the most successful aspect of the Lower Swansea Valley Project has been the tree planting which has transformed the appearance of the valley over the past 20 years. The experimental work to try to grow plants on inhospitable tips was of academic interest, but the findings were not to solve the problem. The toxic smelter wastes have been sandwiched between layers of steel waste to give a chemical environment in which the high pH inhibits the movement of metals. A level plateau has been created in the central part of the valley upon which new industrial units can be established. The present pattern of land use is shown on Fig. 8.8.

The Rising Sun Site

Reclamation of a former area of coal mining and heavy industry in the Ebbw Fach Valley at Blaina, South Wales, illustrates the problems encountered on coalfield sites. The Rising Sun Site occupied 74 ha in a mining valley with a long history of industrial exploitation. Local records indicate that coal mining began just over 200 years ago and industrialization began with the construction of the Nant-y-Glo works in 1811. In 1880, the site was occupied by an industrial complex which included mines, coke ovens, brickworks, limekilns, and the Blaina ironworks. There had been extensive tipping of coal shale on the valley floor and the river had been culverted.

By 1901, the Blaina ironworks had been transformed into a tinplate works and a new ferro–manganese–spiegeleisen works had been constructed on its northern side. The Great Western Railway had built a line up the valley and the various mines and factories were linked by tramways. By 1920 tipping had begun on the western valley side where it continued until 1975 when the last active colliery closed after an underground fire. The locations of these former factories and mines is recorded on successive issues of the Ordnance Survey maps (Fig. 8.9).

On the hill slopes, falls caused by the instability of the Inkerman and Red Ash tips destroyed houses. A row of cottages known as Alma Row was carried away between 1900 and 1920 and another slide in 1927 threatened Colliers Row and caused several properties to be demolished. The last recorded movement of the tips took place in 1959. The valley floor tips were reworked for the usable coal in the

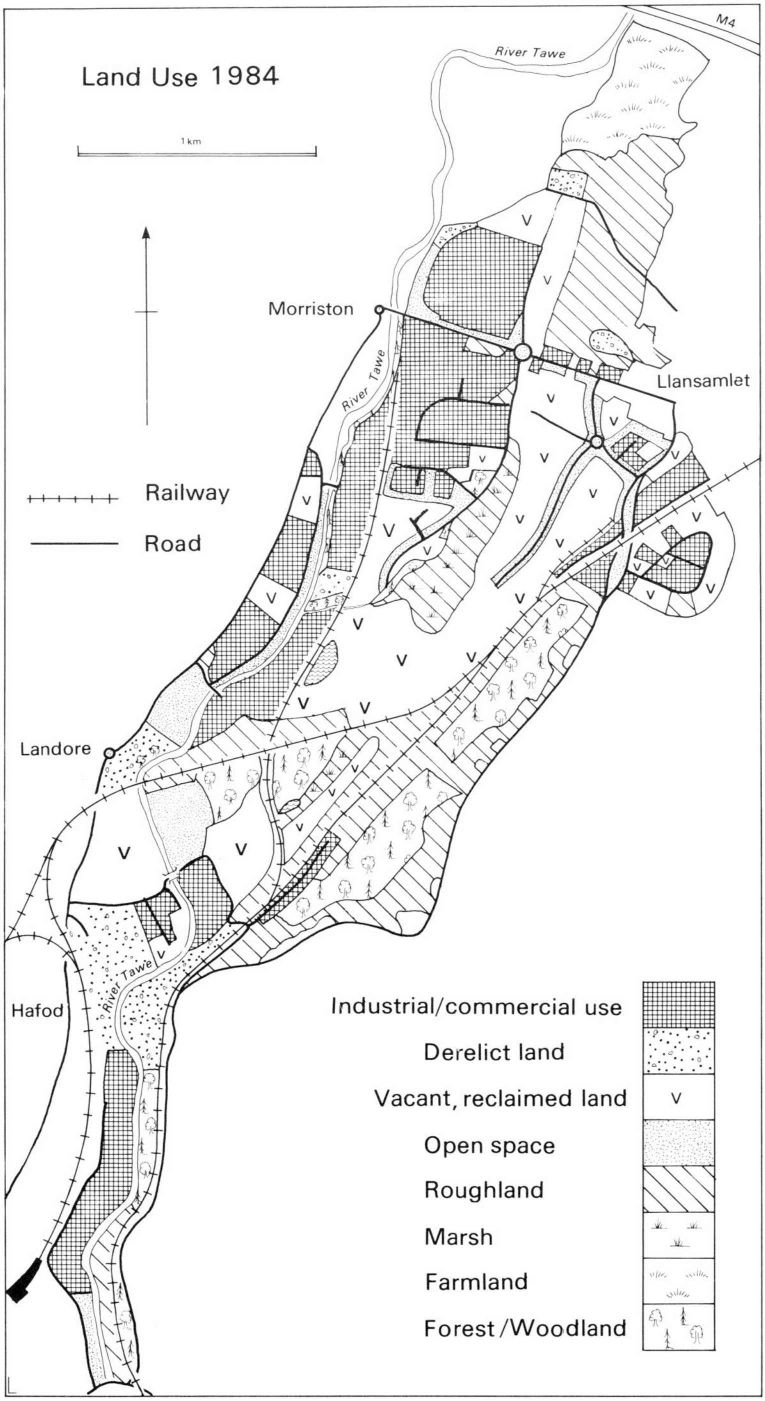

Fig. 8.8. Land use in the Lower Swansea Valley, 1984.

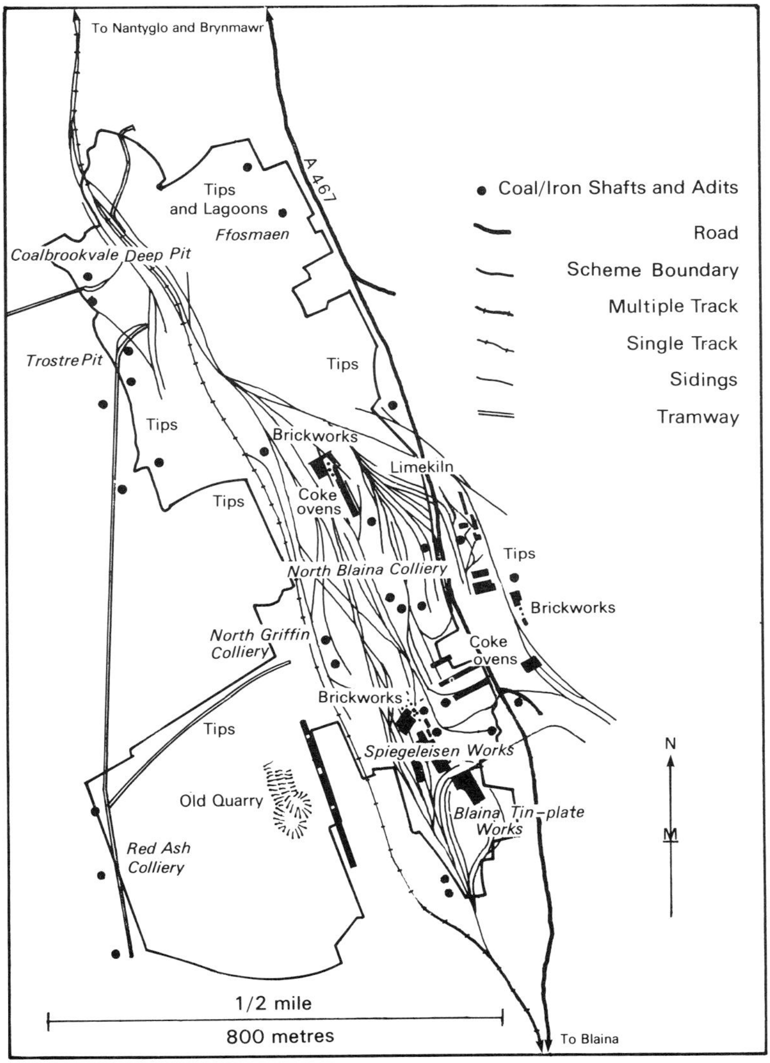

Fig. 8.9. Industrial features of Blaina and Nant-y-Glo (1880–1920). (Based upon the Ordnance Survey Map, with permission of the Controller of HMSO. Crown Copyright.)

early 1970s, leaving numerous slurry lagoons in the northern part of the site. The conditions on the site were reported in a site survey (Robinson, Jones Design Partnership 1981).

> . . . the various industries which once used the area left their mark mainly in the form of extensive and ugly waste tips. They consist of colliery shale, ironstone shale, ironwork slab, slurry from coal washing and other debris which covers the valley floor to such a degree that only a small portion of the original ground surface is exposed anywhere over the site.

In 1973, the Urban District Council of Nant-y-Glo and Blaina asked consultants to prepare a design for rehabilitation and reuse of the site; feasibility reports were received the following year. Reclamation work began in 1977 by subcontractors, administered and supervised by the Robinson, Jones Design Partnership.

Examination of the late nineteenth-century culvert carrying the river revealed that it was not in sufficiently good condition to withstand any further load. This culvert had to be replaced before any earth moving could begin. A new pipe of special strength concrete, 2100 mm in diameter, was laid along nearly 1 km of the former colliery railway line on the valley floor, thus keeping excavation costs to a

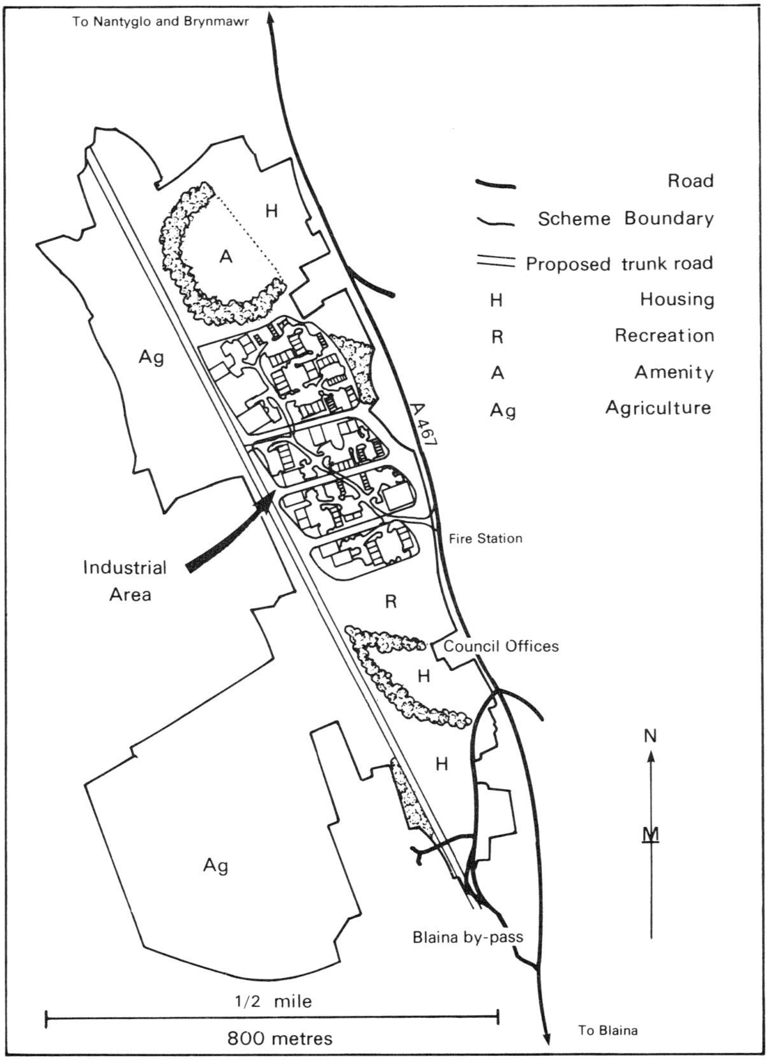

Fig. 8.10. Reclamation scheme for the Rising Sun Site at Blaina and Nant-y-Glo, 1973).

minimum. An existing main foul sewer also had to be replaced, together with subsidiary sewage pipes, and a medium-pressure gas main which crossed the site was realigned alongside the existing main road (A467).

As this was a former mining area, great care was taken to locate former coal and iron ore mineshafts; 28 shafts and several adits were recorded by the National Coal Board. These mines were located and exploratory borings made to check they were completely safe. Where appropriate, shafts were filled, capped, or plugged with a reinforced cement–pfa mixture. This operation involved the injection of nearly 800 tons of cement grouting.

Once these initial tasks of making safe were completed, major earth moving could begin. All derelict buildings and other structures were demolished and the hillside waste tips were regraded, contoured, and excess shale material was brought down on to the valley floor within the site boundaries. Altogether 1.5 million m^3 of shale were moved. The contouring of the hillside re-established the topography prior to tipping, and large areas on the valley floor were levelled for industrial or housing development. An adequate drainage system was installed which conveys excess water to the Ebbw Fach by reinforced concrete channels and culverts. The work is now complete (1984) despite delays caused by subcontractors going into liquidation and the restriction of funding. The proposed land use site following restoration is shown in Fig. 8.10. The total cost of reclamation was £7 m.

Restoration of derelict railway land

In 1841, the Great Western Railway decided to set up its locomotive and wagon works in a field adjacent to the small market town of Swindon, Wiltshire. The works continued to expand from 1843 and by 1921 covered 326 acres and employed 3500 people (Fig. 8.11). However, by the 1960s rationalization of the railway system released for redevelopment the land north of the main line and east of the branch line to Stroud which is now known as the North Star Site.

This site had been occupied by railway sidings and various concrete tanks and foundations remained on the surface. It had been used for 120 years during which time it was covered with waste dumps of partly burnt coal, coke, and ash from the fires of steam engines. Dumping had covered at least 20 acres; the depth of material increased from a few feet thick in the west towards the east where it attained 18 ft, with an average depth of 15 ft. Subsequently, railway sidings were extended on to the area from the adjacent locomotive works. Two other sites, one to the north-east of Swindon station and the other lying north of the main line on the western outskirts of the town, had also been used as ash dumps and contained similar problems (Fig. 8.12).

Early maps of the Ordnance Survey indicate that the North Star Site contained a gasworks in the north-west corner (the gasholders are still in use) and in the centre, a brickworks, which was subsequently covered by the ash dumps. As the geological material below the ash deposits is a firm brown or grey-brown clay overlying a stiff grey Kimmeridge Clay of the Jurassic formation, it is unlikely that there will be excessive leaching of any toxic substances from the site, especially from the gasworks, other than in the superficial layers. The brickworks was probably a relatively small concern and may have left some solid foundations and brick rubble together with some (now infilled) excavations from which the brick clay was obtained. This could cause some variability in ground-water conditions on site.

Situated immediately north of Swindon station, the North Star Site occupies a prime site, relatively close to the town centre. The local authority began redevelopment by constructing a swimming pool complex, ('The Oasis') and the Wiltshire Technical College (Fig. 8.13). In 1973 the tipped material ignited. The underground fire burned vigorously and endangered the Technical College and a hospital laundry on the eastern margin of the site (Hurwitz 1973). As an emergency measure the local authority installed concrete-filled trenches between the burning material and these buildings at a cost of £75 000.

The local authority considered the possibility of allowing the ash inside the barriers to burn itself out

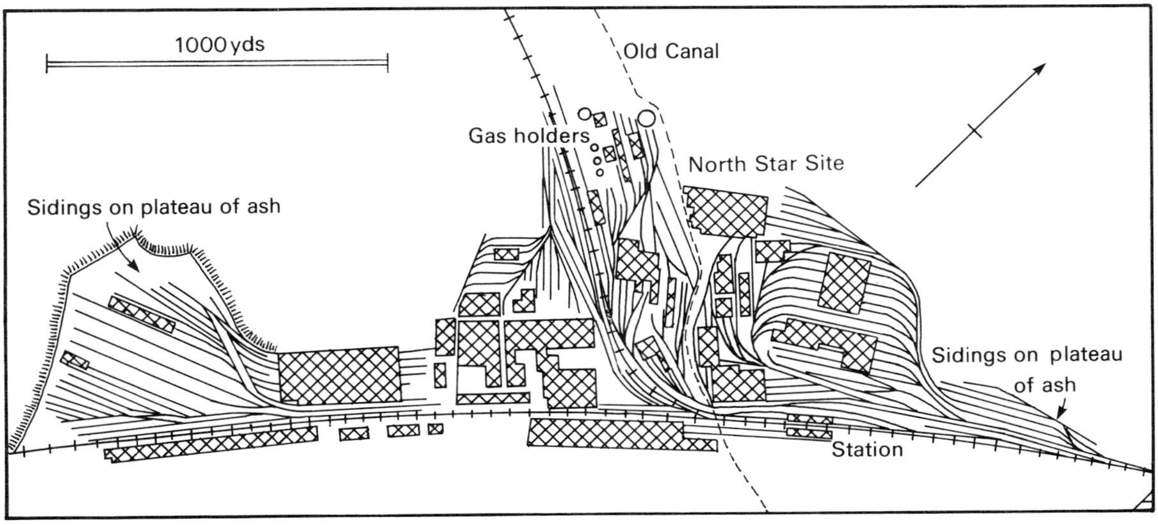

Fig. 8.11. Land occupied by railways, Swindon, *c.* 1956. (Based upon the Ordnance Survey Map, with permission of the Controller of HMSO. Crown Copyright.)

Fig. 8.12. Ash and coal dumps on railway land at Swindon.

but this would have created fumes and dust and sterilized the site for an indefinite period. Also it would have raised problems of public safety since the underground fires left a thin surface crust below which temperatures were described as 'white hot'. Anyone attempting to walk over the burning tip risked breaking the crust and falling into the fire below, which could result in major burns or possibly death. An attempt to use the ash for manufacture of concrete blocks was unsuccessful and so finally the decision was taken to remove it.

Over a period of two years a contractor removed 250 000 tons of ash, but public complaints led to the contractors dampening the ash whilst working and shielding the grading machines which threw dust into the air causing a nuisance to a neighbouring housing area. 120 000 m^3 of the excavated material contained sufficient coal to warrant its shipment to Didcot power station where it was burnt and re-emerged as ash which was used to reclaim gravel workings.

The Science Research Council building has since been erected on the south-east corner of the North Star Site, adjacent to the main railway line. Since the ash was removed from the site, this building is situated at a lower level than the surrounding land.

The firm Norwest Holst Soil Engineering Ltd, investigated the central part of the North Star Site where an L-shaped area of derelict land remains between the Oasis swimming pool and the Science Research Council's headquarters. Twelve boreholes, of 102 mm diameter, uniformly scattered throughout the area, were considered adequate to investigate subsurface conditions. The boreholes were logged and *in situ* compaction tests made in the field. The borehole logs revealed the worked ground of the old brickworks and two boreholes had to be relocated because of substantial unidentified obstructions at depth. Undisturbed samples were obtained at regular depths wherever cohesive materials were encountered. They were tested for moisture content, liquid and plastic limits, organic content, sulphates, pH, CBR, consolidation, and triaxial compression.

As a result of these field and laboratory investigations, the consultants divided the area into three parts which ideally should be developed separately. However, it is likely that the developments proposed (which include shopping or office facilities and multistorey car parking) will straddle their boundaries. As a result of differences between the worked clays and the *in situ* clays beneath, piled foundations were recommended for all clay structures. Where they penetrated the water-table, it was recommended that the piles should be cased and sealed into the clay beneath to prevent chemical attack upon the pile shaft. As the samples revealed that between 9 and 20 per cent combustible material remained,

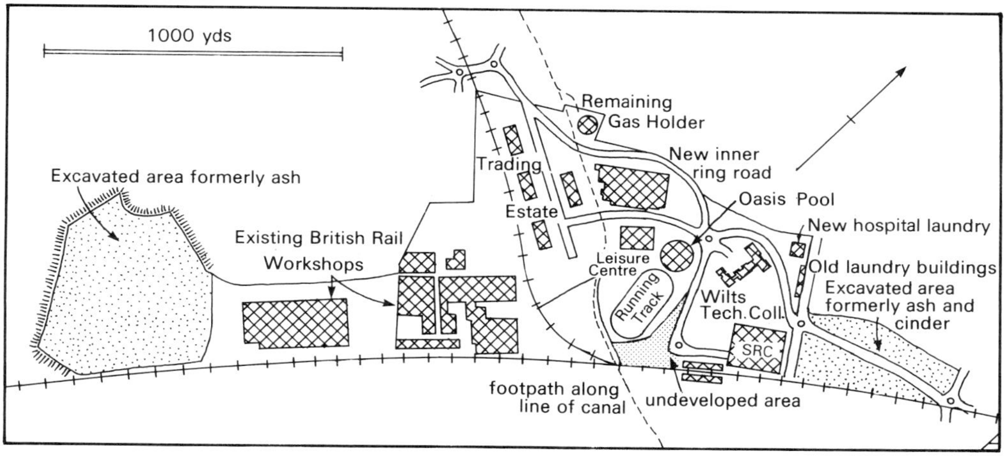

Fig. 8.13. Sketch map of land use in Swindon, 1984.

it was stressed that the water-table should not be lowered, as there would be a fire risk, and that entry of air be prevented.

The line of the old canal has been utilized as a footpath and an industrial estate has been set up between it and the branch line to Stroud. Part of the former gasworks site was also included in this estate and the newly constructed inner ring road has isolated the remaining gasholder from the rest of the redeveloped area. A leisure centre has been built alongside the Oasis swimming pool and a new hospital laundry, adjacent to the premises formerly threatened by the underground fire.

The two other sites, at either end of the Railway holdings in Swindon, also became available for redevelopment. Both areas had been built up and levelled with ashes and waste coal, and railway sidings were laid upon them. Most of this material has now been removed by contractors and, on the eastern site, part of the inner ring has been constructed. The western site has been worked for reclaimable coal and the separation process has left dried slurry and extensive heaps of ash and clinker. This area was used for cleaning and servicing engines so that the slurry and clinker was found to be contaminated with lead and copper; thus although the slurry contained reasonable levels of plant nutrients, its use as a covering medium was precluded.

British Rail has contracted its operations to the central part of the site, and further contraction is expected. A sketch map of the current land use (1984) is given in Fig. 8.13.

London Docklands and Thamesmead

In the 200-year history of London as a port, many changes have taken place but none so rapid or so fundamental as those occurring at the present time.

Moorings were first available along the river frontage at 'legal quays' which had all the disadvantages of a tidal river. Henry VIII set up a Royal Shipyard at Deptford in 1513 and almost 300 years later the first basin docks were constructed further up river, nearer to the City. These were the St Katherine's Dock, the London Dock, and the East and West India Docks, all of which were constructed between 1799 and 1828 for sailing craft. During the nineteenth century the average size of ships using the London Docks increased from less than 200 tons to about 1000 tons. As sail gave way to steam, these docks were deepened and modernized but were still small in size. Continuing growth of trade in the nineteenth century led to the construction of the Royal Victoria Dock in 1855 and the Royal Albert Dock in 1880 (Fig. 8.14). The Tilbury Docks were constructed in 1886.

A succession of mergers between rival companies resulted in the formation of the Port of London Authority (PLA) in 1909. The docks continued operations under the PLA until the mid-1960s, when, as a result of the movement of activity downstream, some of the older and smaller docks were closed. St Katherine's Dock was closed in 1963 and was subsequently bought by the Greater London Council. The East India Dock closed in 1967, the London

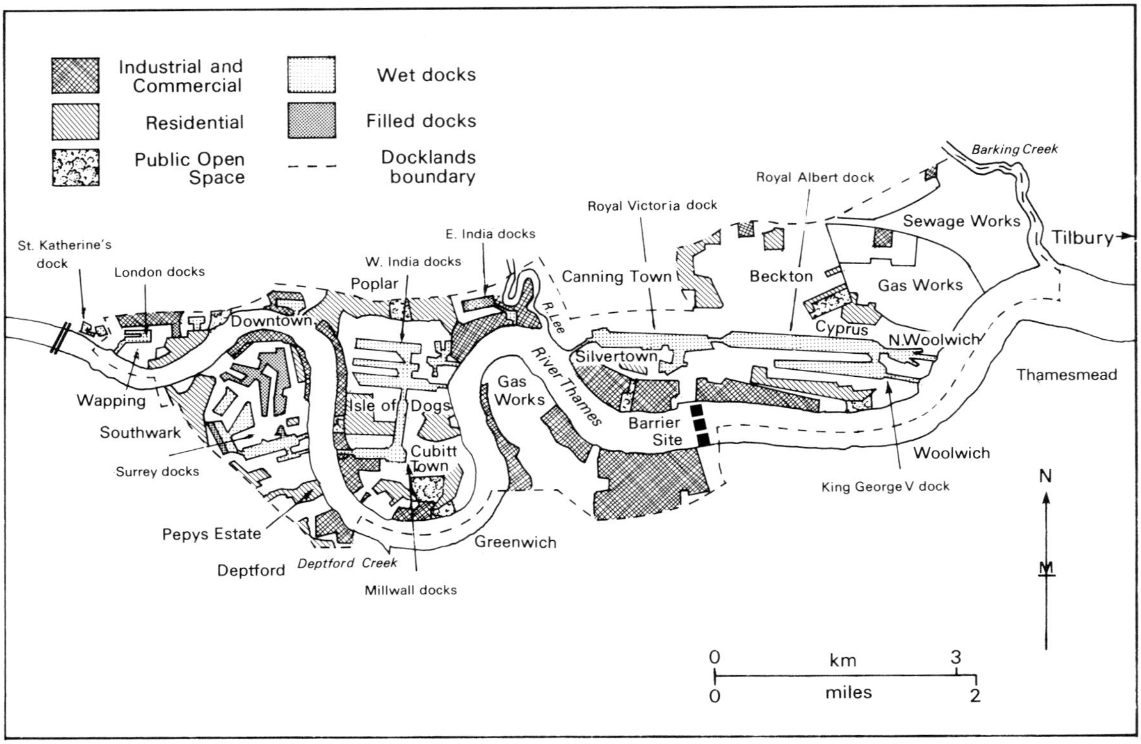

Fig. 8.14. Land use in the London Docklands, 1973 (Travis, Morgan and Partners).

Dock in 1968, and the Surrey Docks in 1970. The Royal Group of Docks ceased to be used for general cargo-handling in 1980 and the India/Millwall Docks in 1981.

Over the same period, the PLA was investing in modern facilities downstream, transforming Tilbury from a relatively minor dock into one of the most modern, comprehensive ports in Britain. There was also investment in deep-water berths downstream, between Greenwich and Gravesend, designed for specific cargoes. Victoria Deep Water Terminal at Greenwich has a £2.5 m. container-handling facility, and Convoys Wharf, Deptford specializes in handling newsprint. The Northfleet Hope deep-water container terminal opened in 1978 adjacent to Tilbury Docks, costing £24 m. A further RO-RO (roll on–roll off) berth opened in 1980 at Erith costing £1 m.

Many ancillary industries have been associated with the docks and often intermixed with dockside activity. After the foundation of the London County Council in 1888, obnoxious dockside industries were forced east of the River Lea, resulting in the industrialization of Canning Town and Silver Town with glue and soap factories. In 1878 Tate built a sugar refinery adjacent to the Royal Victoria Dock and his competitor Lyle did likewise. The two refineries remained in competition until they linked in 1921. Figure 8.15 shows the diversity of industry along the Thames in the early-1960s and indicates the major railways of the Royal Docks. In the sixteenth and seventeenth centuries, Huguenots forced out of France settled in East London bringing their skills in the silk industry. Later other clothing factories were attracted to the area by cheap female and migrant labour.

During the nineteenth century ropemakers, sailmakers, and shipwrights concerned with wooden sailing ships lost their jobs as the steam-driven iron ships required different skills in their construction and maintenance. The London dockland area had a wide range of small industries surrounded by the homes of the artisans. Such small labour-intensive industries proved unable to meet the challenges of the last two decades and as the trade of the docks declined so did their fortunes. Severe job losses

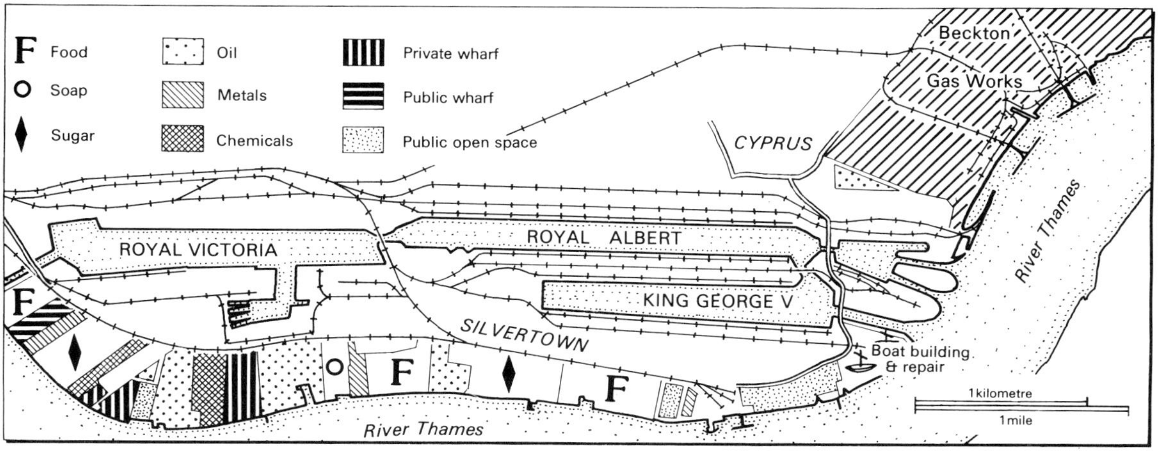

Fig. 8.15. Diversity of industry along the Thames near the Royal Docks in 1964 (Castle and Stone 1964).

dominated the docklands between 1961 and 1971. In Tower Hamlets alone, over 10 000 jobs were lost in the clothing, furniture, food, drink, and tobacco industries, and a further 9300 in the transport industry. Overall, Dockland lost 82 750 manufacturing jobs and 23 000 jobs in transport including the docks. This continued until redevelopment began during the 1980s. In 1971 there were 2.76 square miles of derelict land and an estimated 4 square miles in 1981. This led to '. . . the greatest rebuilding opportunity since the Great Fire of 1666' (Vigars 1976).

Several vaulted warehouses and cellars of architectural merit formerly used for storing wine, rum, brandy, and tobacco surround the London Docks. These could be used for clubs, discotheques, or cellars for dwellings constructed above. The arches are too close for underground car parking. Industrial buildings in good condition with some architectural merit occupy 2 500 000 ft^2; a further 500 000 ft^2 of buildings in poor condition occur which have no historic value.

There are 19 000 dwellings on the docklands at present, of which 2000 are in such poor condition that they should be demolished, a further 4000 require immediate attention, and others will need it in the next few years. Proposals are in hand to build 23 000 houses at Beckton and Isle of Dogs at a Density of 70–100 rooms per acre, but between the areas of new housing at Beckton and Barking Creek, a considerable area is to be retained for industrial use, such as the active Beckton sewage works and the derelict Beckton gasworks site.

Planning for reclamation

Consultants were commissioned by the DoE and GLC in 1971 to study the London Docklands. The PLA was not a part to the study but co-operated fully with the team. The study was concerned with 2200 ha (8.5 square miles) lying between Wapping and Barking which in 1971 contained a population of 55 000 and provided 60 000 jobs. A report (Anon 1973) was submitted which described five main steps for improvement:

1. Identifying the problems of the area and measuring them;
2. Devising a list of national development possibilities;
3. Comparing the possibilities and producing a shortlist;
4. Working up the development possibilities of the shortlist in more detail and considering their possible staging and timing;
5. Setting out the social and financial impact of each.

As it was not known what pressures might operate in the future to close further docks, this aspect was left open. The five options studied were referred to as: The Waterside Plan; The Thames Park Plan; The East End Consolidated Plan; Europa Plan; City New Town Plan.

The Waterside Plan This envisaged the use of the docks as an attractive focus for an improved urban environment for a population of nearly 74 000 by 1991. Except for the establishment of new parks at

Poplar and Silvertown and an educational sports centre at Beckton, nearly all the land was to be devoted to housing. The retained waterways would provide moorings for small craft and a marina for larger boats was proposed. A maritime museum was to be established in the Poplar basin.

TOTAL COSTS: estimated at £666m.

The Thames Park Scheme This gave extra emphasis to the development of open spaces by the creation of three major parks: in Surrey Docks; the Isle of Dogs; and the Beckton area. It envisaged a population of 50 000 with expansion and relocation of commercial and industrial premises. It proposed a concentration of leisure activities in the Beckton area with marinas, caravan site, holiday hotel, and safari park.

TOTAL COST: £728m.

East End Consolidated Plan This aimed at a higher population and attempted to reverse the declining employment prospects. Seventy-five per cent of the housing would be owned by local authorities and rented, and the remaining 25 per cent would comprise private accommodation.

TOTAL COST: £768m.

Europa Plan This envisaged a more up-market version of the East End Consolidated Plan with a similar population but only 33 per cent of the housing to be owned by local authorities. A rapid transit system was to be developed and a chain of lakes, new parks, and other specialized areas to improve the environment.

TOTAL COST: £894m.

City New Town Plan This aimed to house 106 000 by 1991 following the pattern set in the New Towns. A new centre was to be created on the Isle of Dogs with shopping facilities and offices. A parkland was to be created to the east of the centre.

TOTAL COST: £948m. (Lunch 1973.)

These plans were put on public exhibition just before a change in local government and were promptly scrapped by the incoming political party. A new initiative by the Docklands Joint Committee in 1974 undertook a consultative approach with the local people which had been lacking previously. By 1976, a 'Strategy for Docklands' had been prepared (Docklands Joint Committee 1976).

This envisaged that the population of the area would increase from 55 000 to 120 000 and that 23 000 new houses should be built, 35 per cent owned by local authorities, 20 per cent owner-occupied, and the remaining 45 per cent on an equity-sharing basis (part rent/part own) with housing associations. A key proposal was the building of the River Line, an extension of the London Underground railway from Charing Cross through to Fenchurch Street, Millwall, Blackwell, Victoria Dock, terminating at the new township of Thamesmead. This plan rejected major shopping centres and hypermarkets in favour of local shopping facilities. Six hundred and twenty acres were allocated for industry and 360 acres for public open space.

COST: £1143m with £900m private investment.

Although no final design has appeared, nor the finance for it, parts of the Dockland area have been rehabilitated. Over 100 houses were built in 1976 and building continues to the present day. The 826-roomed Tower Hotel, a marina, and a floating maritime museum have been constructed at St Katherine's Dock. A former warehouse has been reconstructed to provide premises for the London World Trade Centre (Leith 1979). The West India Docks now specializes in the bulk import of wine and recently became the new home of Billingsgate fish market, transferred from its original site between the Tower and London Bridge (Copeman 1979). Small workshops have been set up in buildings around the West India Docks under the Enterprise Zone scheme and 130 workshop units (Cannon workshops) ranging from 160 ft^2 to 1700 ft^2 have provided 500 new jobs (Warren 1982). Several national newspapers have transferred their activities from Fleet Street to Wapping and the Isle of Dogs, where new premises house modern technological equipment.

The Docklands Development Corporation was set up in 1981 funded by central government to recreate a thriving, prosperous region from the former port. In 1982, this was reinforced by the designation of the Isle of Dogs Enterprise Zone, specifically to encourage new industry into the area.

Specific problems

Docks The landscape of dockland is flat, with few natural features, so buildings and trees must be used to create focal points. Tree planting is envisaged as an important part of rehabilitation along the Thames and in the new parks. As land ownership is by various bodies and persons, the Docklands Joint Committee requested that a joint board for land assembly in docklands be set up under the Community Land Act 1975, so that a unified development can take place.

Derelict docks can either be filled or developed as water areas for amenity or recreation. In the case of the London dockland, the Surrey Docks (except the Greenland Docks) have been filled already and on the north bank of the Thames, 69 acres of land have been reclaimed from the East India Docks, leaving 111 acres of open water. Proposals for the remaining docks all involve amenity or recreational areas (Fig. 8.16).

At present, the Docklands Strategy Plan has 1525 acres available for redevelopment, but only 400 acres could be used immediately, as the remainder needs preparation before development could commence. A large drainage scheme for 800 acres was undertaken by the Thames Water Authority at Beckton in 1978. Two major problem industries occupied large areas at Beckton—a sewage works and a gasworks. The sewage works is to continue operations and requires additional land, but the change to natural gas has released much of the land formerly occupied by the gasworks. An industrial estate is being developed on 65 acres of this land, including an area of tips known as Beckton Alps. The problem of development on such sites were introduced in Chapter 3 and a description of the covering system used can be found on p. 94.

South of the river, on the infilled Surrey Docks, a large commercial development (Trade Mart) is under discussion. £150 000 has been spent on river frontage work and it is planned that a central park in the Surrey Docks area, will be created with extensive sports facilities. Ambitious plans are being discussed for a £35m housing scheme with water-based recreational facilities for the Greenland and South Docks areas.

One of the proposed developments at the Royal Docks site is the possibility of a short take-off and landing (STOL) airport. It has been proved possible to operate STOL aircraft from this site for short-haul flights. A potential market of over a million passengers is estimated and development costs are thought to be in the region of £2.5m.

Gasworks sites The Beckton gasworks originally covered 179 ha but, since conversion to natural gas, only the gasholders have been used. Some land had already been redeveloped in 1971 for a mineral water factory, a further 39 ha of the by-products site was scheduled for redevelopment, and a substantial area designated for an extension of the Beckton sewage works.

The consultants' report (Travis, Morgan and Partners 1973) about the Beckton Gasworks site stated that '. . . the whole area of the works has been raised from marsh level with ash and clinker varying from 1 m to 3 m in depth and with up to 0.6 m of spent oxide in parts'. The tipped materials forming the Beckton 'Alps' include spent oxide, gas lime, boiler ash, and clinker. The spent oxides contain iron oxide and up to 40 per cent sulphur; the gas lime consists of calcium carbonate with sulphides removed from the gas. Both contain cyanides. On the site there were lagoons of tar, still soft below a surface crust, and rubble from demolished buildings. Sulphuric acid, ammonia, and tar acid products were manufactured in the by-products plant. Surface spillages of tar were evident in this area and it is likely there were spillages of other by-products. Tar pits were present and a ditch surrounding the perimeter of the site had to be pumped out regularly as it is contaminated.

The consultants proposed to remove the tarry material from the tar pits and to reduce the slope of the sides of the tipped materials, covering them with 1.3 m of London Clay and 30 cm of fresh soil.

To determine sulphur content and acidity, a series

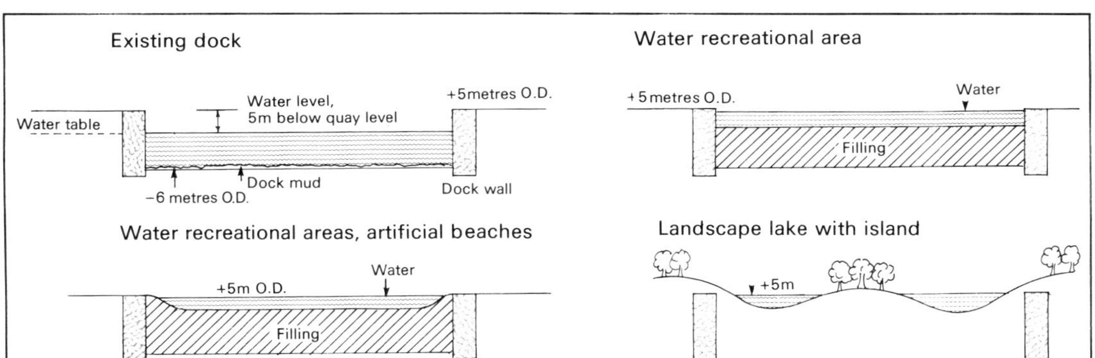

Fig. 8.16. Possible solutions for the use of derelict docks (Travis, Morgan and Partners 1973).

of test boreholes over the whole site was recommended, but tests revealed that no abnormal attack on concrete would take place, except in the vicinity of the waste acid plant. Thirty millimetres of soil was thought to be sufficient for grass and trees over most of the site, but a footnote indicated that the site would not be suitable for private gardens or allotments. All buildings on the site were to be piled, or rested on concrete rafts.

On the south side of the river, the East Greenwich gasworks covered 98 ha. A car park (8 ha) and storage areas without contamination problems were released in 1972 for redevelopment. Before 1880, part of the site was an ordnance factory but after that date the gasworks spread over the whole site. The by-products works had been built alongside a former dry dock, which was used as a water reservoir. East of this reservoir the land had been raised by dumping spent oxides and sulphides, both containing cyanides. There had been spillages of creosote and sulphuric acid. The area is now heavily contaminated, partly through bomb damage during the Second World War. Seepages of tar and creosote leak into the Blackwell Tunnel which lies beneath the site.

Beckton Sewage Works This is one of London's principal treatment plants and purifies 40 per cent of the city's sewage. The effluent water is discharged to the Thames, and the sludge dumped in the North Sea. The sewage works are being extended, so former sewage works land is unlikely to become available for redevelopment. Additional ground is required for an emergency sludge lagoon and further adminstrative buildings, including a store. A refuse transfer station is planned for part of the site and a broad swathe across the area is earmarked for a new ringway road which will intersect the A13 and carry traffic over a new planned Thames crossing.

Thamesmead Major contamination problems have been encountered south of the Thames at Thamesmead, the site of the former Royal Arsenal at Woolwich. The site was acquired by the Greater London Council piecemeal during the 1960s and a start was made on the construction of Thamesmead, a new township for 60 000 people. In 1975, heavily contaminated ground was encountered, there was a chemical fire, and tanks of phenolic liquids were discovered below the ground surface. Building had to be suspended because of the risk to workers on site and the potential risks for future site occupants if these contaminants were not removed (Lowe 1980) (Fig. 8.17).

A working party, which included scientific and medical advisers, investigated the problem in conjunction with the Thamesmead Derelict Land Remedial Unit. A derelict land survey confirmed that a wide range of industrial activities had taken place on site, including the manufacture of gas, metal-working, the manufacture and destruction of explosives, and the dumping of paint and other industrial wastes. A careful examination was made of each subdivision of the site in a preliminary survey to plan the sampling strategy. The basic sampling interval was 100 m reducing to 50 m or even 10 m where there was evidence of past industrial activity

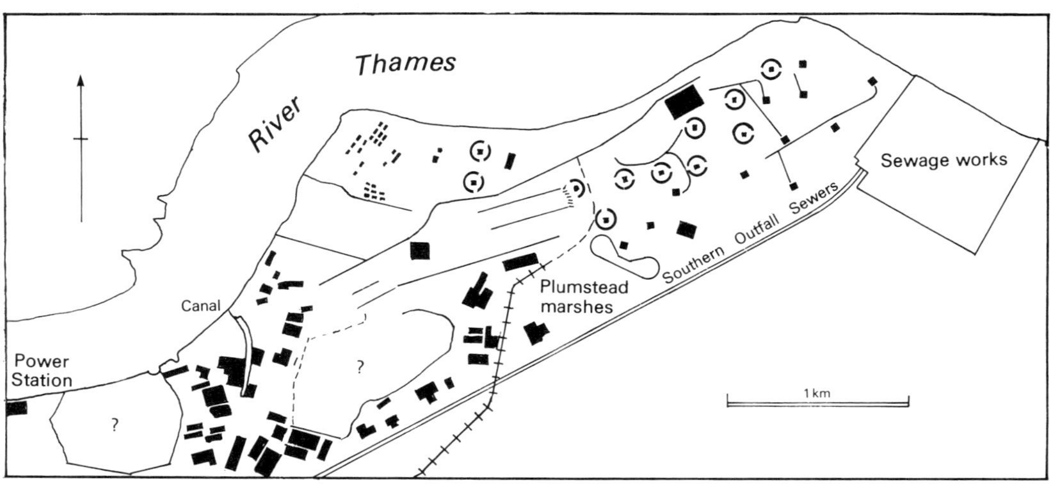

Fig. 8.17. The site of the former Woolwich Arsenal (Ordnance Survey 1:25 000 series). (Based upon the Ordnance Survey Map, with permission of the Controller of HMSO. Crown Copyright.)

which may have caused contamination. The planned sampling grid was drawn-up on a 1:5000 map. Each sample point was marked on the ground by a numbered stake, and then a pit was dug with an excavator under the supervision of an officer of the Derelict Land Remedial Unit, and samples were taken to a depth of 3 m. Log sheets were completed and colour photographs were taken to document the site materials fully (Anon 1978).

The effect of site contamination on the immediate health of site workers and on the long-term environment was evaluated from the analyses. The level of contaminants, their calorific value, the depth of ground water and its condition, phytotoxicity, likelihood of methane formation, the underground corrosion of concrete structures, and the means of dealing with the contaminants were all assessed against the proposed development.

The planners adopted a 'loose-fit' approach to site development, incorporating many of the existing site features into the plan for Thamesmead. It retains some of the lakes and canals which convey surface water from upland areas across the site and drain into the Thames, and some of the circular mounds and moats around buildings formerly used for storage of high explosives. The lead-enriched firing range had been safely buried. It is intended to place 800 mm hydraulically dredged sea sand plus 300 mm of topsoil over contaminants to give a safe covering and growth medium. The exceptionally polluted area of the former arsenal gasworks is saturated by phenolic liquids, so this site has been assigned to public open space. It will be covered with 1 m of broken concrete and rubble, blinded with shingle, and capped with 300 mm of clay, the surface of which will be shaped to shed water to at least 1 m below the soil surface at the periphery. No trees will be planted, but the area will be laid out as playing fields.

It is a fortunate coincidence that the construction of flood defences for London required a new river wall along the northern boundary of the Thamesmead site. The space between the old river bank and the new wall has been authorized as a site to receive contaminated material, which therefore need not leave the site. When this filling is complete, it will be capped with clean sand as described.

So far the redevelopment of the site is half completed and the Derelict Land Remedial Unit has succeeded in keeping restoration work ahead of building operations (Fig. 8.18), working in close collaboration with the DoE's Central Unit on Environmental Pollution.

Estimated costs of restoration (1975)

On average, demolition in the docklands of London costs £12 000 per ha at 1975 prices. Additional costs of site clearance, such as removal of buried objects and the provision of foundations for intensive site redevelopment, has ranged from £8200 to £47 000 but the overall average additionial cost is between £8200 and £9900 per ha. Reclamation of the actual dock areas allowing for infilling and extra piling, is estimated to cost between £27 000 and £130 000 per ha. Replacement of the river walls costs an average of £1 200 000 per km but in this case renewal was required for flood defences so reclamation did not bear the costs.

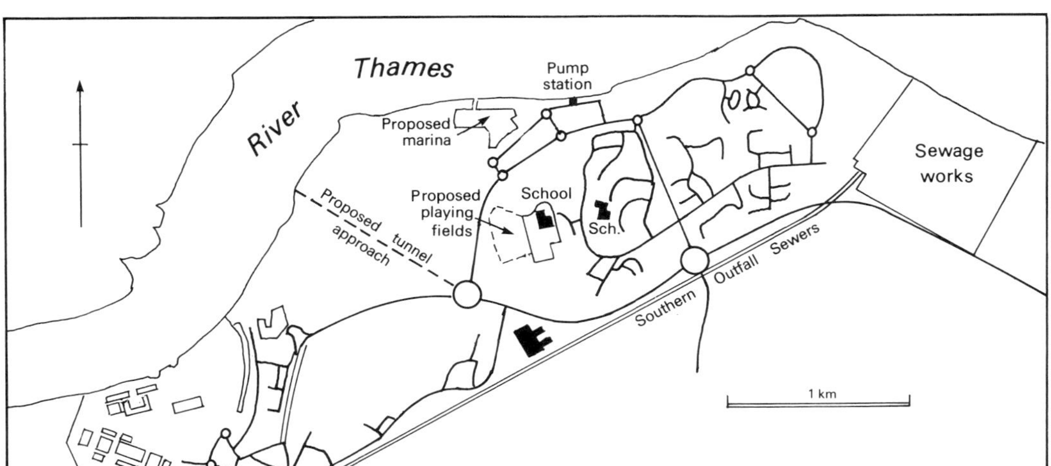

Fig. 8.18. The development of Thamesmead from derelict, contaminated land, 1982.

References

Aldag, R. W. and Strzyszcz, Z. (1980). Inorganic and organic nitrogen compounds in carbonaceous phyllosilicates and spoils with regard to forest reclamation. *Reclamation Review* **3**, 69–73.

Allen, S. E., Grimshaw, A. M., Parhura, J. A., and Quarmly, C. (1974). *Chemical analysis of ecological materials.* Blackwell Scientific Publications, Oxford.

Alloway, B. J. and Morgan, H. (1986). The behaviour and availability of Cd, Ni and Pb in polluted soils. In *Contaminated soil* (ed. J. W. Assink and W. J. van den Brink) pp. 101–15. Martinus Nijhoff, Dordrecht.

American Conference of Government Industrial Hygienists (1974). *Threshold values limit for metals in workroom air.* Cincinnati, Ohio.

Anon (1973). Dockland study. *Port of London* **48**, (3) No. 568, 81–2.

—— (1978). London tackles industrial waste hazard. *International Construction* (May) 18–22.

Aspinwall, R. (1975). Hydrological and hydrogeological assessment of landfill sites. *Solid Wastes* **65**, 390–403.

—— (1976). Hydrogeology in land reclamation. In *Land reclamation.* pp. 355–67. Thurrock Borough Council, Grays, Essex.

Avery, B. W. and Bascomb, C. L. (1974). *Soil survey laboratory methods.* Soil Survey, Harpenden.

Baker, D. H. and Bhappu, R. B. (1974). Specific environmental problems associated with the processing of minerals. In *Extraction of minerals and energy: today's dilemmas.* (ed. R. A. Defu). Ann Arbor Science, Michigan.

Baker, J. M. (1971). Oil and salt marsh soil. In *The ecological effects of oil pollution on littoral communities.* Proceedings of Institute of Petroleum Seminar.

Barber, E. G. (1975). *Win back the acres: the treatment and cultivation of pfa surfaces.* Central Electricity Generating Board, London.

Barltrop, D. (1986). Evaluation of cadmium exposure from contaminated soil. In *Contaminated soil* (ed. J. W. Assink and W. J. van den Brink) pp. 169–81. Martinus Nijhoff, Dordrecht.

Barnes, G. E. (1986). The effects of groundwater flow pattern on the concentration of soluble sulphates. In *Contaminated soil* (ed. J. W. Assink and W. J. van den Brink) pp. 115–21. Martinus Nijhoff, Dordrecht.

Barnes, M. (1977). *Examples of the CESMM.* Institute of Civil Engineers, London.

Barr J. (1969). *Derelict Britain.* Pelican, Penguin Books, Harmondsworth.

Barry, D. L. (1986). Hazards from methane on contaminated sites. In *Contaminated soil* (ed. J. W. Assink and W. J. van den Brink) pp. 537–41. Martinus Nijhoff, Dordrecht.

Beaver, S. H. (1946). *Report on derelict land in the Black Country.* Mimeograph from Ministry of Town and Country Planning, London.

—— (1969). An appraisal of the problem. In *Proceedings of Derelict Land Symposium.* Iliffe Science and Technology Publications, Leeds.

Beavington, F. (1975). Heavy metal contamination of vegetables and soil in domestic gardens around a smelting complex. *Environmental Pollution* **9**, 211–18.

Beever, P. F. (1986). Assessment of fire hazard in contaminated land. In *Contaminated soil* (ed. J. W. Assink and W. J. van den Brink) pp. 515–23. Martinus Nijhoff, Dordrecht.

Bell, R. M., Gildon, A., and Parry, G. D. R. (1983). Sampling strategy and data interpretation for site investigation of contaminated land. In *Reclamation of former iron and steel works sites* (ed. G. P. Doubleday). Durham and Cumbria County Councils, Durham.

Binns, W. O. and Fourt, D. F. (1984). Reclamation of surface workings for trees. I. Landforms and cultivation. *Aboriculture Research Note* no. 37/84/SSS. Forestry Commission, Alice Holt, Farnham.

Blood, J. W., Jackson, H., and Ormrod, J. E. (1961). Restoration to agriculture of land worked for ironstone and levelled without the addition of topsoil. *Experimental Husbandry* **6**, 90–9.

Bocock, K. L. (1981). *Radionuclides in terrestrial ecosystems.* National Environment Research Council, Institute of Terrestrial Ecology.

Bowen, H. J. M. (1979). Erosional chemistry of the elements. Academic Press, New York.

Bradshaw, A. D. and Chadwick, M. J. (1980). *The restoration of land.* Studies in Ecology, Vol. 6, Blackwell Scientific Publications, Oxford.

——, Dancer, W. S., Handley, J. F., and Sheldon, J. C. (1975). The biology of land revegetation and the reclamation of the china clay wastes of Cornwall. In *The ecology of resource degradation and renewal.* (ed. M. J. Chadwick and G. T. Goodman) pp. 363–85. Blackwell, Oxford.

Brent-Jones, E. (1976). The agricultural restoration of open cast coal sites in Great Britain. In *Land reclamation,* pp. 243–6. Thurrock Borough Council, Grays, Essex.

—— (1983). Opencast coal restoration 1983. In *Reclamation 83,* pp. 57–81, Proceedings of International Land Reclamation Conference. Industrial Seminars Ltd, Tunbridge Wells.

Bridges, E. M. (1967). New land for old in the East Midlands. *East Midland Geographer* **4**, 143–53.

—— (1969). Eroded soils of the Lower Swansea Valley. *Journal of Soil Science* **20**, 236–45.

——, Chase, D. S., and Wainwright, S. J. (1979). Soil and plant investigations since 1967. In *Dealing with dereliction* (ed. R. D. F. Bromley and G. Humphrys) pp. 23–43. University College of Swansea, Swansea.

——, ——, and —— (1981). Distribution of copper, lead, zinc, cadmium and nickel in plants, superficial organic layers and mineral soils of the Lower Swansea Valley. In *The productivity of restored land.* Land Decade Educational Council, London.

Brink, A. B. A., Partridge, T. C., and Williams, A. A. B. (1982). *Soil Survey for Engineering.* Monographs on Soil Survey, Oxford University Press, Oxford.

British Standards Institute (BSI). *Code of practice for site investigations,* BS 5930. BSI, London.

—— (1983). *Draft British Standard code of practice for the identificaton and investigaton of contaminated land.* EPC/47. BSI, London.

—— (1985). *Use of industrial by-products and waste materials in building and civil engineering.* BSI 6543, London.

Bromley, J. (1980). Gas generation and the problems it creates in land reclamation. In *Reclamation of contaminated land.* Society of the Chemical Industry, London.

Bromley, R. D. F. and Humphrys, G. (eds.) (1979). *Dealing with dereliction.* University College of Swansea, Swansea.

—— and Morgan, R. H. (1983). *Change and industrial redevelopment in the Lower Swansea Valley,* University College of Swansea, Swansea.

Brown, D. W. (1976). The possible use of industrial waste and its transportation for reclamation purposes. In *Land reclamation,* pp. 145–59. Thurrock Borough Council, Grays, Essex.

Building Research Establishment (BRE) (1981). Concrete in sulphate heavy soils and groundwater. *BRE Digest,* no. 250. DoE, London.

Burrows, J. (1978). Vacant urban land—a continuing crisis. *The Planner* **64**, 7–9.

Bush, P. W. (1969). Spoiled lands to the south-east of Leeds. In *Proceedings of the Derelict Land Symposium.* Iliffe Science and Technology Publications, Leeds.

Cairns, J., Dickson, K. L., and Horricks, E. E. (eds.) (1977). *Recovery and restoration of damaged ecosystems.* University of Virginia Press, Charlottesville.

Campbell, J. M. (1983). Burning colliery waste and dust pollution: the reclamation challenge at Gorseinon. In *Reclamation 83,* pp. 586–95. Industrial Seminars Ltd., Tunbridge Wells.

Carpenter, R. J. (1986). Redevelopment of land contaminated by methane gas: the problems and some remedial techniques. In *Contaminated soil* (ed. J. W. Assink and W. J. van den Brink) pp. 747–59. Martinus Nijhoff, Dordrecht.

Castle, D. M. and Stone, G. (1964). The London Docks. In *Guide to London excursions* (ed. K. M. Clayton). 20th International Geographical Congress, London.

Central Electricity Generating Board (CEGB) (n.d.). *Land Reclamation—Peterborough.* CEGB (Midland Region).

Central Statistical Office (1980). *Standard industrial classification.* HMSO, London.

Chadwick, M. J. and Goodman, G. T. (eds.) (1975). *The ecology and resource degradation and renewal.* 15th Symposium of the British Ecological Society, Blackwell Scientific Publications, Oxford.

Chase, D. S. (1978). *Botanical and economic aspects of revegetating part of the Lower Swansea Valley.* Unpublished Ph.D. thesis, University of Wales, Swansea.

—— and Wainwright, S. J. (1983). The vertical distribution of copper, zinc, lead in weathered tips of copper smelter waste in the Lower Swansea Valley. *Environmental Pollution* (Series B), **5**, 133–46.

Childs, K. A. (1985). In-ground barriers and hydraulic measures. In *Contaminated land: reclamation and treatment* (ed. M. A. Smith), pp. 145–82. Plenum, London.

Clark, I. S. (1977). Financial control and contract procedure. In *Landscape reclamation practice* (ed. B. Hackett), pp. 173–205. IPC Science and Technology Press, Guildford.

Clausen, H. T. (1976). Aspects of mine dump reclamation on the Witwatersrand. In *Land reclamation,* pp. 121–35. Thurrock Borough Council, Grays, Essex.

Coleman, A. and Shaw, J. (1980). *Field mapping manual.* Second Land Utilisation Survey of Britain, King's College, London.

Collins, W. G. and Bush, P. W. (1969*a*). The definition and classification of derelict land. *Journal of the Town Planning Institute* **55**, 111–15.

—— and —— (1969*b*). The use of aerial photographs in the study of derelict land—an air photo key. *Journal of the Town Planning Institute* **55**, 246–55.

CONCAWE, (1983). *A field guide to inland oil spill clean-up techniques, Report No. 10/83.* Roepers Drukkerig, The Hague.

Connolly, C. J. and Veitch, P. J. (1983). Computer aided landscape design. In *Reclamation of former iron and steel works sites* (ed. G. P. Doubleday). Durham and Cumbria Country Councils, Durham.

Cope, C. B., Fuller, W. H., and Willetts, S. L. (1983). *The scientific management of hazardous wastes.* Cambridge University Press, Cambridge.

Copeman, S. (1979). West India dock development. *Port of London* **54** (3), 101–3.

Cornwell, S. M. and Stone, E. L. (1968). Availability of nitrogen to plants in acid coal mine spoils. *Nature* **217**, 768–9.

Coulson, M. G. and Bridges, E. M. (1984). The remote sensing of contaminated land. *International Journal of Remote Sensing* **5**, 659–69.

—— and —— (1986). Site assessment and monitoring of contaminants by airborne multi-spectral scanner. In *Contaminated soil* (ed. J. W. Assink and W. J. van den Brink), pp. 365–78. Martinus Nijhoff, Dordrecht.

Cowie, W. D. (1981). The reuse of contaminated land. *Proceedings 48th Annual Conference National Society for Clean Air,* Brighton.

Craze, B. (1977). Investigations into the revegetation problems of Captain's Flat mining area. *Journal of Soil Conservation of New South Wales* **33**, 190–9.

Culbard, E. B. and Essex, J. (1983). The chalk cliffs of Thurrock: a land-use planning case study. In *Reclamation 83,* Proceedings of International Land Reclamation Conference. Industrial Seminars Ltd, Tunbridge Wells.

Dahl, T. O. (1986). Occidental Chemical Company at Lathrop, California, a groundwater/soil contamination problem and a solution. In *Contaminated soil* (ed. J. W. Assink and W. J. van den Brink), pp. 793–807. Martinus Nijhoff, Dordrecht.

Davies, B. E. (1977). Heavy metal pollution of British agricultural soils with special reference to the role of lead and copper mining. In *Proceedings of International Seminar on Soil Environment and Fertility Management in Intensive Agriculture,* Tokyo.

—— (1978). Plant available lead and other metals in British garden soils. *The Science of the Total Environment* **9**, 243–62.

—— (1983). *Halkyn Mountain Project Report.* Welsh Office, Cardiff.

—— and Lewin, J. (1974). Chronosequences in alluvial soils with special reference to historic lead pollution in Cardiganshire, Wales. *Environmental Pollution* **6**, 49–57.

—— and Roberts, L. J. (1978). The distribution of heavy metal contaminated soils in north-east Clwyd, Wales. *Water, Air and Soil Pollution* **9**, 507–18.

Davis, G. (1978). Oil Shale. In *Reclamation of drastically disturbed lands* (ed. F. W. Schaller and P. Sutton). American Society of Agronomy, Madison, Wisconsin.

Davis, R. (1982). Quoted in *Farmers Weekly* 30 May 1982.

Davis, R. D. and Beckett, P. H. T. (1978). Upper critical levels of toxic elements in plants. *New Phytol* **80** (1), 23–32.

Dean, F. E. and Goalby, B. B. (1980). Successful redevelopment of gasworks sites. In *Reclamation of contaminated land.* Society of the Chemical Industry, London.

Docklands Joint Committee (1976). *London strategic plan.* Docklands Development Team.

DoE (1971). *Report of the working party on refuse disposal.* HMSO, London.

—— (1974). *Survey of derelict and despoiled land: guideline notes.* HMSO, London.

—— (1975). *Survey of derelict and despoiled land in England, 1974,* DoE/PRM3/801/252. HMSO, London.

—— (1976*a*). *Reclamation, treatment and disposal of wastes.* Waste Management Paper No. 1, HMSO, London.

—— (1976*b*). *Waste disposal surveys.* Waste Management Paper No. 2, HMSO, London.

—— (1976*c*). *Guidelines for the preparation of a waste disposal plan.* Waste Management Paper No. 3, HMSO, London.

—— (1976*d*). *The licensing of waste disposal sites.* Waste Management Paper No. 4, HMSO, London.

—— (1976*e*). *Relationship between waste disposal authorities and print industry.* Waste Management Paper No. 5, HMSO, London.

—— (1976*f*). *Polychlorinated biphenyl (PCB) wastes.* Waste Management Paper No. 6, HMSO, London.

—— (1976*g*). *Mineral oil wastes.* Waste Management Paper No. 7, HMSO, London.

—— (1976*h*). *Heat treatment cyanide wastes.* Waste Management Paper No. 8, HMSO, London.

—— (1976*i*). *Halogenated hydrocarbon solvent wastes for cleansing processes.* Waste Management Paper No. 9, HMSO, London.

—— (1976*j*). *Metal finishing wastes.* Waste Management Paper No. 11, HMSO, London.

—— (1977). *Derelict land.* Circular 1977/17. London.

—— (1977*a*). *Asbestos health hazards and precautions.* HMSO, London.

—— (1977*b*). *Report of the working party on the disposal of sewage sludge to land.* DoE and National Water Council, London.

—— (1977*c*). *Mercury-bearing wastes.* Waste Management Paper No. 12, HMSO, London.

—— (1977*d*). *Tarry and distillation wastes and other chemical-based residues.* Waste Management Paper No. 13, HMSO, London.

—— (1977*e*). *Solvent waste (excluding halogenated hydrocarbons).* Waste Management Paper No. 14, HMSO, London.

—— (1977*f*). *Halogenated organic waste.* Waste Management Paper No. 15, HMSO, London.

—— (1978*a*). *Wood preserving wastes.* Waste Management Paper No. 16, HMSO, London.

—— (1978*b*). *Wastes from tanning, leather dressing and fellmongering.* Waste Management Paper No. 17, HMSO, London.

—— (1978*c*). *Asbestos waste—a technical memorandum on arisings and disposal including a code of practice.* Waste Management Paper No. 18, HMSO, London.

—— (1978*d*). *Wastes from the manufacture of pharmaceuticals, toiletries and cosmetics.* Waste Management Paper No. 19, HMSO, London.

—— (1978*e*). *Joint agricultural land restoration experiments.* HMSO, London.

—— (1980). *Pesticide wastes.* Waste Management Paper No. 21, HMSO, London.

—— (1981). *Special wastes: a technical memorandum providing guidance on their definition.* Waste Management Paper No. 23, HMSO, London.

—— (1984). *Survey of derelict land in England in 1982.* DoE, London.

—— and National Water Council (1979). Report of the Working Party on the Disposal of Sewage Sludge to Land. DoE, London.

Doubleday, G. P. (1971). Soil forming materials: their nature and assessment. In *Landscape reclamation,* Vol. 1, Ch. 8, pp. 70–83. IPC Science and Technology Press, Guildford.

Downing, M. F. (1971). Landform design. In *Landscape reclamation.* Vol. 1, Ch. 4, pp. 32–42. IPC Science and Technology Press, Guildford.

—— (1977). Survey information. In *Landscape reclamation practice* (ed. B. Hackett) pp. 17–36. IPC Science and Technology Press, Guildford.

Dumbleton, M. J. (1980). Historical investigation of site use. In *Reclamation of contaminated land.* Society of the Chemical Industry, London.

—— and West, G. (1976). *Preliminary source of information for site investigation in Britain.* Report LR403, Transport and Road Research Laboratory, DoE.

Edelman, Th. and de Bruin, M. (1986). Background values of 32 elements in Dutch topsoils, determined with non-destructive neutron activation analysis. In *Contaminated soil* (ed. J. W. Assink and W. J. van den Brink) pp. 89–101. Martinus Nijhoff, Dordrecht.

Edwards, P. R. and Pumphrey, N. W. J. (1980). Preparations for reclamation of a lagoon containing canal dredgings. In *Reclamation of contaminated land.* Society of the Chemical Industry, London.

Egginton, M. J. (1983*a*). The town that died of dioxin poisoning. *The Observer* 16 January.

—— (1983*b*). The town America wrote off for cash. *The Observer* 27 February.

Elias, J. (1979). The role of the Welsh Development Agency. In *Dealing with dereliction* (ed. R. D. F. Bromley and G. Humphrys) pp. 207–19. University College of Swansea, Swansea.

Ellis, R. and Adams, R. S. (1961). Contamination of soils by petroleum hydrocarbons. *Advances in Agronomy* **13**, 197–216.

Gemmell, R. P. (1973). Revegetation of land polluted by a chromate smelter 1. Chemical factors causing substrate toxicity in chromate smelter waste. *Environmental Pollution* **5**, 181–97.

—— (1977). *Colonization of industrial wasteland.* Institute of Biology's Studies in Biology, No. 80. Arnold, London.

Goodman, G. T. (1974). Ecological aspects in the reclamation of derelict land. In *Conservation in practice* (ed. A. Warren and F. B. Goldschmith) pp. 251–64. Wiley, Chichester.

—— (1979). Airborne heavy metal pollution. In *Dealing with dereliction. The redevelopment of the Lower Swansea Valley* (ed. R. D. F. Brownley and G. Humphrys) pp. 57–93. University College of Swansea, Swansea.

—— and Roberts, T. M. (1971). Plants and soils as indicators of metals in air. *Nature London* **231**, 287–92.

——, Smith, S., and Inskip, M. J. (1975). Moss bag study of atmospheric pollution—South Wales. Moss bags as indicators of airborne metals—an evaluation. In *Report of a collaborative study on certain elements in air, soil, plants, animals and humans in the Swansea–Neath–Port Talbot areas together with a report on a moss bag study of atmospheric pollution across South Wales,* pp. 267–364. Welsh Office, Cardiff.

Greenwood, E. F. and Gemmell, R. P. (1978). Derelict industrial land as a habitat for rare plants in South and West Lancs. *Watsonia* **12**, 33–40.

Gutt, W., Nixon, P. J., Smith, M. A., Harrison, W. H., and Russell, A. D. (1974). *A survey of the locations, disposal and prospective uses of the major industrial by-products and waste materials.* CP 19/74, Building Research Establishment, DoE, London.

Hackett, B. (ed.) (1977). *Landscape reclamation practice.* IPC Science and Technology Press, Guildford.

Haines, R. (1980). *A comparison of the extent of heavy metal contamination on vacant land sites and background urban levels.* Research Note 11, JURUE, University of Aston, Birmingham.

—— (1981). *Contaminated sites in the West Midlands County. A prospective survey.* JURUE, University of Aston, Birmingham.

Hance, R. J. (Ed.) (1980). *Interactions between herbicides and the soil.* Academic Press, New York/London.

Hatayama, H. (1986). Site investigations: a review of current methods and techniques. In *Contaminated soil* (ed. J. W. Assink and W. J. van den Brink) pp. 315–25. Martin Nijhoff, Dordrecht.

Health and Safety Commission (1985). The protection of persons against ionising radiation arising from any work activity Approved Code of Practice. *The ionising radiations regulations.* HMSO, London.

Heeps, K. D. and Pike, E. R. (1980). Reclamation of a disused sewage farm. In *Reclamation of contaminated land.* Society of the Chemical Industry, London.

Hesse, P. R. (1971). *A textbook of soil chemical analysis.* John Murray, London.

Hewitt, A. K. J. and McRae, S. G. (1986). The effects of gases emitted from landfills on soils and crops. In *Contaminated soil* (ed. J. W. Assink and W. J. van den Brink) pp. 251–5. Martinus Nijhoff, Dordrecht.

Hilton, K. J. (Ed.) (1967). *The Lower Swansea Valley project.* Longman, London.

Huntjens, J. L. M., de Potter, H., and Barendrecht, J. (1986). The degradation of oil to soil. In *Contaminated soil* (ed. J. W. Assink and W. J. van den Brink) pp. 121–5. Martinus Nijhoff, Dordrecht.

Hurwitz, S. (1973). Death lurks in North Star's glowing depths. *Swindon Evening Advertiser* 22 October 1973.

ICRCL (1978*a*). *Redevelopment of contaminated land, general notes.* ICRCL 3/78, DoE, London.

—— (1978*b*). *Notes on the redevelopment of landfill sites.* ICRCL 17/78, DoE, London.

—— (1979*a*). *Redevelopment of gasworks sites.* ICRCL 18/79, DoE, London.

—— (1979*b*). *Progress report of the Interdepartmental Committee on the Redevelopment of Contaminated Land.* ICRCL 19/79, DoE, London.
—— (1979*c*). *Notes on sewage works and farms.* ICRCL 23/79, DoE, London.
—— (1980). *Notes on the redevelopment of scrapyards and similar sites.* ICRCL 42/80, DoE, London.
—— (1983). *Guidance on the assessment and redevelopment of contaminated land.* ICRCL 59/83, DoE, London.
—— (1984). *Notes on the fire hazards of contaminated land.* ICRCL 61/84, DoE, London.
—— (1985). *Asbestos on contaminated sites.* ICRCL 64/85, DoE, London.
Jackson, M. L. (1958). *Soil chemical analysis.* Constable, London.
Jennett, J. C. and Wixson, B. G. (1972). Problems in lead mining waste control. *Journal of the Water Pollution Control Federation* **44** (11), 2103–10.
Jobling, J. (1981). Afforestation of colliery spoil. In *The productivity of restored land.* Land Decade Educational Council, London.
John, D. G. and Davies, D. R. (1976). Some hydrogeological aspects of waste disposal at landfill sites. In *Land reclamation*, pp. 369–77. Thurrock Borough Council, Grays, Essex.
Johnson, A. and King, W. H. (1947). *A manual of specifications and quantities for civil engineers.* English Universities Press, London.
Johnson, M. S., McNeilly, T., and Putwain, P. D. (1977). Revegetation of metalliferous mine spoil contaminated by lead and zinc. *Environmental Pollution* **12**, 261–77.
Jonas, F. (1973). Reclamation of areas damaged by mining activity in Czechoslovakia. In *Ecology and reclamation of devastated land* (ed. R. J. Hutnik and G. Davis). Gordon and Breach, New York.
Jones, A. K., Johnson, M. S., Bell, R. M., and Bradshaw, A. D. (1980). Biological aspects of the treatment of heavy metal contaminated land for housing development schemes. In *Reclamation of contaminated land.* Society of the Chemical Industry, London.
Jones, H. I. I. (1979). Engineering problems and their solution. In *Dealing with dereliction* (ed. R. D. F. Bromley and G. Humphrys) pp. 191–9. University College of Swansea, Swansea.
JURUE (1982). *Background levels of heavy metal soil contamination in Walsall.* University of Aston, Birmingham.
Kelly, R. T. (1980). Site investigation and material problems. In *Reclamation of contaminated land.* Society of the Chemical Industry, London.
Kennet Committee (1939). *Report of the Committee on the Restoration of Land Affected by Ironstone Working.* HMSO, London.
Knabe, W. (1973). Development and application of the DOMSDORF ameliorative treatment of toxic spoil banks of lignite opencast mines in Germany. In *Ecology and reclamation of devastated land* (ed. R. J. Hutnik and G. Davis). Gordon and Breach, New York.
Kooper, W. F. and Mangnus, G. A. M. (1986). Sampling and analysis in contaminated site investigations: impediments and provisional guidelines in the Netherlands. In *Contaminated soil* (ed. J. W. Assink and W. J. van den Brink) pp. 325–37. Martinus Nijhoff, Dordrecht.
Land use Consultants (1970). Stoke on Trent reclamation programme. *Institute of Landscape Architects Journal* **90**, 14–22.
Leith, D. (1979). The new face of St Katherine's Dock. *Port of London* **54** (2) No. 609, 54–7.
Leigh, D. (1982). Death-fluid was hushed up. *Observer* 14 March.
Lindfors, L. G. (1980). Reclamation of the site of a herbicide factory. In *Reclamation of contaminated land.* Society of the Chemical Industry, London.
Loch, J. P. G., Kool, H. J., Lagas, P., and Verheul, J. H. A. M. (1986). Removal and retention of volatile chlorinated hydrocarbons in the soils' insaturated zone. In *Contaminated soil* (ed. J. W. Assink and W. J. van den Brink) pp. 63–79. Martinus Nijhoff, Dordrecht.
Lord, D. W. (1983). The proposed British standard code for the investigation of contaminated sites. In *Reclamation 83*, pp. 176–87. Industrial Seminars Ltd, Tunbridge Wells.
Lowe, G. W. (1980). GLC development at Thamesmead, investigation and reclamation of contaminated land. In *Reclamation of contaminated land.* Society of the Chemical Industry, London.
Lunch, J. (1973). Dockland study. *Port of London* **48** (4) No. 569, 124–5.
Mackenzie, D. (1985). Dioxin: still looking for the bodies. *New Scientist* 26 September, p. 23.
Mather, J. D. and Bromley, J. (1976). Research into leachate generation and attenuation at landfill sites. In *Land reclamation*, pp. 377–401. Thurrock Borough Council, Grays, Essex.
McCarthy, M. J. (1980). Reclamation of a refuse tip for open space and housing development Paper B8. In *Reclamation of contaminated land.* Society of the Chemical Industry, London.
McRae, S. G. (1981). The productivity of restored gravel workings. In *The productivity of restored land.* Land Decade Educational Council, London.
—— (1983). The soil scientist's contribution to quarry design and reclamation. In *Reclamation 83*, pp. 29–36. Proceedings of International Land Reclamation Conference. Industrial Seminars Ltd, Tunbridge Wells.
Mellanby, K. (1980). *The biology of pollution*, 2nd edn. Studies in Biology No. 38, Arnold, London.
Meller, B. B. (1979). The financial commitment of the Swansea City Council. In *Dealing with dereliction* (ed. R. D. F. Bromley and G. Humphyrs) pp. 165–81. University College of Swansea, Swansea.
Michaels, L. and Chissick, S. S. (1979). *Asbestos, Vol. I. Properties, applications and hazards.* Wiley Interscience, Chichester.
Miller, J. B. (1983). The hazards of unstable working faces,

tipped material and silt lagoons in quarries. In *Reclamation 83*, pp. 531–9. Proceedings of International Land Reclamation Conference. Industrial Seminars Ltd, Tunbridge Wells.

Ministry of Agriculture, Fisheries and Food (MAFF) (1981). *Pesticides safety precaution schemes.* Pesticides Branch, MAFF, London.

Ministry of Housing and Local Government (1966). *Circular 59/66.* MHLG, London.

Ministry of Housing, Planning and the Environment (Netherlands) (1983). *Assessing the severity of a case of soil contamination in the Netherlands.* Mimeograph.

Moen, J. E. T., Cornet, J. P., and Evers. C. W. A. (1986). Soil protection and remedial actions: criteria for decision making and standardization of requirements. In *Contaminated soil* (ed. J. W. Assink and W. J. van den Brink) pp. 441–9. Martinus Nijhoff, Dordrecht.

Muir, G. D. (1977). *Hazards in the chemical laboratory.* The Chemical Society, London.

Netherton, D. N. and Tollin, B. I. (1981). Reclamation of the Beckton Alps and adjacent areas. *Public Health Engineer* **10**, 202–21.

Observer (1984). Soviet missiles 'unsafe', 24 June.

Ottaway, I. H. (1980). *The biochemistry of pollution.* Studies in Biology No. 123, Arnold, London.

Ottevanger, W. (1986). Human cadmium intake in a contaminated region. In *Contaminated soil* (ed. J. W. Assink and W. J. van den Brink) pp. 247–51. Martinus Nijhoff, Dordrecht.

Oxenham, J. R. (1966). *Reclaiming derelict land.* Faber, London.

—— (1967). Land reclamation. *Town Planning Institute Journal* **53** (8), 344–7.

Page, R. A. and Swires-Hennessy, E. (1986). A safe level for lead in soil and dust. In *Contaminated soil* (ed. J. W. Assink and W. J. van den Brink) pp. 233–47. Martinus Nijhoff, Dordrecht.

Pankhurst, E. S. (1973). The effects of natural gas on trees and other vegetation. In *Aspects of current use and misuse of soil resources*, pp. 116–30. Welsh Soils Discussion Group Report 14.

Parry, G. D. R. and Bell, R. M. (1983). The impact of landfill gas on open space and agricultural schemes. In *Reclamation 83*, pp. 510–15, Proceedings of International Land Reclamation Conference. Industrial Seminars Ltd, Tunbridge Wells.

—— and —— (1985). Covering systems. In *Contaminated land: treatment and reclamation* (ed. M. A. Smith) pp. 113–33. Plenum, New York/London.

Patterson, J. W. (1979). Industry sources and control of metals. In *Management and control of heavy metals in the environment.* London.

Pickering, W. F. (1977). *Pollution evaluation.* Marcel Dekker, New York.

Plice, M. J. (1948). Some effects of crude petroleum on soil fertility. *Proceedings of the Soil Science Society of America* **13**, 413–16.

Reinhold, G., Kirsch, J., Hurtig, H.-W., and Flothmann, D. (1986). Remote sensing by electromagnetic induction (EMI), magnetometer and ground penetrating radar. In *Contaminated soils* (ed. J. W. Assink and W. J. van den Brink) pp. 378–89. Martinus Nijhoff, Dordrecht.

Rimmer, G. (1976). Putting solid ground beneath Corby's feet. *Surveyor* **147**, 24–5.

Robinson, Jones Design Partnership (1981). *Rising Sun Reclamation Scheme.* Mimeograph.

Rulkens, W. H., Assink, J. W., and van Gemert, W. J. Th. (1985). On-site processing of contaminated soil. In *Contaminated land: treatment and reclamation* (ed. M. A. Smith) pp. 37–91. Plenum, New York/London.

Russell, E. J. (Russell, W.) (1973). *Soil conditions and plant growth.* 10th edn. Longman, London.

Sanning, D. E. (1985). *In-situ* treatment. In *Contaminated land: treatment and reclamation* (ed. M. A. Smith) pp. 91–113. Plenum, New York/London.

Schmehl, W. R. and McCaslin, B. D. (1973). Some properties of spent oil shale significant to plant growth. In *Ecology and restoration of devastated land* (ed. R. J. Hutnik and G. Davis). Gordon and Breach, New York.

Schuuring, C. (1981). Dutch dumps. *Nature* **289**, 340.

Schwendinger, R. B. (1968). Reclamation of soil contaminated with oil. *Journal of the Institute of Petroleum* **54**, 182–97.

Skinner, B. J. (1969). *Earth resources.* Prentice-Hall, New Jersey.

Skinner, R. J. (1973). Contamination of soil by oil and other chemicals. *Welsh Soils Discussion Group* **14**, 131–41.

Smith, L. C. (1980). Hazards to immediate workers on site. In *Reclamation of contaminated land.* Society of Chemical Industry, London.

Smith, M. A. (Ed.) (1985). *Contaminated land: treatment and reclamation.* Plenum Press, New York/London.

—— and Ellis, A. C. (1986). An investigation into methods used to assess gasworks sites for reclamation. *Reclamation and Revegetation Research* **4**, 183–209.

Spellerberg, I. F. (1981). *Ecological evaluation for conservation.* Studies in Biology No. 133, Arnold, London.

Steere, N. V. (1967). *Handbook of laboratory safety.* Chemical Rubber Company (CRC), Cleveland, Ohio.

Stevens, R. (1976). *Planning control over mineral working.* HMSO, London.

Stief, K. (1985). The long-term effectiveness of remedial measures. In *Contaminated land: treatment and reclamation* (ed. M. A. Smith). Plenum, New York/London.

Street, H. E. and Goodman, G. T. (1967). Revegetation techniques in the Lower Swansea Valley. In *The Lower Swansea Valley report* (ed. K. J. Hilton) pp. 71–110. Longman, London.

Sumner Report (1978). *Co-operation programme of research on the behaviour of hazardous wastes in landfill sites.* HMSO, London.

Thomas, G. H. (1983). Properties of iron and steel slags. In *Reclamation of former iron and steelworks sites* (ed. G. P. Doubleday). Durham and Cumbria County Councils.

Thomson, G. H. and Aldridge, J. A. (1983). London docklands—problems: annotated with their development with special reference to the Surrey Docks scheme. In *Reclamation 83*, pp. 124–33, Proceedings of International Land Reclamation Conference. Industrial Seminars Ltd, Tunbridge Wells.

Travis, Morgan and Partners (1973). *London docklands.* Mimeograph.

Treshow, M. (1976). *The human environment.* McGraw-Hill, New York.

Tucker, A. (1983). Seveso poison turns up in French village. *Guardian*, 20 May.

Vigars, L. (1976). The redevelopment of London's docklands. *Port of London* **51** (598), 70–2.

Wallwork, K. L. (1974). *Derelict land: origins and prospects of a land-use problem.* David and Charles, Newton Abbot.

Warren, C. (1982). Bringing new life to the docklands. *Port of London* **57** (3) No. 622, 90–3.

Waterhouse, P. S. (1980). The sampling and analysis of polluted sites. In *Reclamation of contaminated land.* Society of the Chemical Industry, London.

Welsh Council (1976). *Derelict land in Wales.* HMSO, London.

Welsh Office (1975). *Report of a collaborative study on certain elements in air, soil, plants, animals and humans in the Swansea/Neath/Port Talbot Area, together with a report on a moss bag Study of Atmospheric Pollution across South Wales.* Crown Copyright, Welsh Office, Cardiff.

—— (1985). *Survey of land for mineral workings.* Cardiff.

Westlake, G. E. (1984). Hazards to birds from pesticides in soils. In *Organic agrochemicals in the soil environment*, pp. 58–67, Welsh Soils Discussion Group, Report 25.

Weston, R. L., Gadgil, P. D., Salter, B. R., and Goodman, G. T. (1965). Problems of revegetation in the Lower Swansea Valley, an area of extensive industrial dereliction. In *Ecology and the industrial society* (ed. G. T. Goodman, R. W. Edwards, and J. M. Lambert) pp. 297–327. Blackwell, Oxford.

Whitcut, M. (1981). Derelict land and reclamation problems in Telford. *Chartered Land Surveyor/Chartered Mineral Surveyor* **3** (2), 3–26.

Wilson, D. C. and Stevens, C. (1981). *Problems arising from the redevelopment of gasworks and similar sites.* AERE Report R 10366, DoE, London.

Wood, A. A. and Ross, A. H. (1980). Reclamation of agricultural land used for disposal of sewage and sewage sludge. In *Reclamation of contaminated land.* Society of the Chemical Industry, London.

World Health Organization (1982). *Rapid assessment of sources of air, water and land pollution.* Offset Publication No. 62, WHO, Geneva.

Wright, P. S. (1981). *Soil survey of the Ffos Las site.* Soil Survey of England and Wales, Rothamsted.

Yland, M. W. F. (1986). Contamination from a coal-tar processing chemical industry: investigations and remedial actions. In *Contaminated soil* (ed. J. W. Assink and W. J. van den Brink). pp. 831–49. Martinus Nijhoff, Dordrecht.

Yost, K. J. (1984). Cadmium, the environment and human health: an overview. *Experimentia* **40**, 157–64.

Young, C. P. and Barber, C. (1980). Contaminated land—effects on water resources. In *Reclamation of contaminated land.* Society of the Chemical Industry, London.

Index